A Course in Item Response Theory and Modeling with Stata

® Copyright © 2018 StataCorp LLC
All rights reserved. First edition 2018

Published by Stata Press, 4905 Lakeway Drive, College Station, Texas 77845
Typeset in LaTeX 2_ε
Printed in the United States of America

10 9 8 7 6 5 4 3 2 1

Print ISBN-10: 1-59718-266-4
Print ISBN-13: 978-1-59718-266-9
ePub ISBN-10: 1-59718-267-2
ePub ISBN-13: 978-1-59718-267-6
Mobi ISBN-10: 1-59718-268-0
Mobi ISBN-13: 978-1-59718-268-3

Library of Congress Control Number: 2017957532

Contents

Figures

Tables

Preface

More than half a century ago, a far-reaching revolution started in behavioral, educational, and social measurement, which to date has also had an enormous impact on a host of other disciplines ranging from biomedicine to marketing. At that time, item response theory (IRT) began finding its way into these sciences. In many respects, IRT quickly showed important benefits relative to the then conventional approach for developing measuring instruments that was based on "classical" procedures.

Since the 1950s and the influential early work by F. Lord in IRT (for example, Lord [1952, 1953]), more than 60 years have passed that have been filled with major methodological advances in this field and more generally in behavioral and social measurement. The intervening decades have also witnessed an explosion of interest in IRT and item response modeling (IRM) across those disciplines as well as the clinical, biomedical, marketing, business, communication, and cognate sciences. These developments are also a convincing testament to the rich opportunities that this measurement approach offers to empirical scholars interested in assessing various latent constructs, traits, abilities, dimensions, or variables, as well as their interrelationships. The latent variables are only indirectly measurable, however, through their presumed manifestations in observed behavior. This is in particular possible via use of multiple indicators or multi-item measuring instruments, which have become highly popular in the behavioral and social sciences and well beyond them.

This book has been conceptualized mainly as an introductory to intermediate level discussion of IRT and IRM. To aid in the presentation, the book uses the software package Stata. This package offers, in addition to its recently developed IRT command, many and decisive benefits of general purpose statistical analysis and modeling software. After discussing fundamental concepts and relationships of special relevance to IRT, its applications in practical settings with Stata are illustrated using examples from the educational, behavioral, and social sciences. These examples can be readily "translated", however, to similar utilizations of IRM also in the clinical, biomedical, business, marketing, and related disciplines.

We find that several features set our book apart from others currently available in the IRT field. One is that unlike a substantial number of treatments of IRT (in particular older ones), we capitalize on the diverse connections of this field to the comprehensive methodology of latent variable modeling as well as related applied statistics frameworks. In many aspects, it would be fair to view this book as predominantly handling IRT and IRM, somewhat informally stated, as part of the latent variable modeling methodology. In particular, the discussion throughout the book benefits as often as possible from the conceptual relationships between IRT and factor analysis, specifically, nonlinear factor analysis. Relatedly, whenever applicable, the important links between IRM and other statistical modeling approaches are also pointed out, such as the generalized linear model and especially logistic regression. Another distinguishing feature of the book is that it is free of misconceptions about and incorrect treatments of classical test theory (Zimmerman 1975). Regrettably, they can still be found in some measurement literature and inhibit significantly in our opinion progress in social and behavioral measurement. In addition, these misconceptions contribute to a compartmentalization approach that seems to have been at times followed especially when disseminating or teaching IRT in circles with limited or no prior familiarity with it. That approach and resulting restrictive focus of interest is in our view highly undesirable. The reason is that such an approach has the potential of creating long-term disservice to the cause of behavioral and social measurement. In this connection, we would also like to point out that unlike many previous treatments, this book presents its discussion and developments without any juxtaposition of IRT to classical test theory. This is because IRT does not need this kind of "comparison" and related misconceptions to convince scholars of what it can deliver under its assumptions (see also Raykov and Marcoulides [2016b]). A third characteristic of the book is that it demonstrates the straightforward, user-friendly, and highly effective Stata applications for IRT modeling. We hope to gain in this way many new enthusiasts for this methodological field as well as IRT software across these and related disciplines. Last but not least, our book aims to provide a coherent discussion of IRT and IRM independently of software. The goal is thereby to highlight as often as possible and in as much detail as deemed necessary important concepts and relationships in IRT before moving on to its applications. This was necessary because in our experience, many individuals seem to find some features of this modeling approach more difficult to deal with and use to their advantage than what may be seen as "conventional" applied statistical concepts and relationships. These features include in particular the inherent nonlinearity in studied item-trait relationships as well as produced estimates (predictions) of individual trait levels and measures of uncertainty associated with them. That difficulty in appreciating characteristic properties of IRT may have arguably resulted from insufficient discussion and clarification of them in some alternative accounts or presentations.

This book could be considered aimed mainly at students and researchers with limited or no prior exposure to IRT. However, we are confident that it will also be of interest to more advanced students and scientists who are already familiar with IRT, in particular owing to the above mentioned features in which the book does not overlap with the majority of others available in this field. In addition, a main goal was to enable readers to pursue subsequently more advanced studies of this comprehensive and complex methodological field and its applications in empirical research, as well as to follow more technically oriented literature on IRT and IRM. Relatedly, even though the book uses primarily examples stemming from the educational and behavioral sciences, their treatment, as well as more generally of this measurement field, allows essentially straightforward applications of the used methods and procedures also in other social science settings. These include the clinical, nursing, psychiatry, biomedicine, criminology, organizational, marketing, and business disciplines (for example, Raykov and Calantone [2014]).

Our book has been influenced substantially by deeply enriching interactions with a number of colleagues over the past years. Special thanks are due to K. L. MacDonald and R. Raciborski for their many instructive inputs on Stata uses and applications in relation to examples used in the book, as well as on IRT and its empirical utilizations in more general terms. The importance of the contributions also of Y. Marchenko and C. Huber cannot be overstated, who provided instrumental support during our work on the book. We are especially indebted to C. Huber for helpful comments and criticism on an earlier version, which contributed markedly to its improvement. His assistance during the book-production phase was similarly invaluable, as was that of the book editor and the production assistant. We also wish to express our particular gratitude to M. D. Reckase, B. O. Muthén, D. M. Dimitrov, M. Edwards, C. Lewis, R. Steyer, S. Rabe-Hesketh, A. Skrondal, and A. Maydeu-Olivares for valuable discussions on IRT and IRM and related applied statistics and measurement approaches. We are similarly thankful to C. Falk, R. J. Wirth, N. Waller, R. Bowles, I. Moustaki, R. D. Bock, S. H. C. duToit, G. T. M. Hult, and J. Jackson for insightful discussions on IRT applications and software. We are also grateful to a number of our students in the courses we taught over the last few years who offered very useful feedback on the lecture notes we first developed for them, from which this book emerged. Last but not least, we are more than indebted to our families for their continued support in lots of ways that cannot be counted. The first author is indebted to Albena and Anna; the second author is indebted to Laura and Katerina.

Tenko Raykov and George A. Marcoulides

Notation and typography

In this book, we assume that you are somewhat familiar with Stata: you know how to input data, use previously created datasets, create new variables, fit regression models, and perform similar analytic activities.

We designed this book for you to learn by doing, so we expect you to read its chapters 2 and 6 through 12 while at a computer trying to use the sequences of commands contained in the book to replicate our results. In this way, you will be able also to generalize these sequences to suit your own needs.

We use the `typewriter` font to refer to Stata commands, syntax, and variables. A "dot" prompt followed by a command indicates that you can type verbatim what is displayed after the dot (in context) to replicate the results in the book.

The data we use in this book are freely available for you to download, using a net-aware Stata, from the Stata Press website, http://www.stata-press.com. In fact, when we introduce new datasets, we load them into Stata the same way that you would. For example,

```
. use http://www.stata-press.com/data/cirtms/lsat.dta
```

Try it. To download the datasets and do-files to your computer, type the following commands:

```
. net from http://www.stata-press.com/data/cirtms/
. net describe cirtms
. net get cirtms
```

1 What is item response theory and item response modeling?

1.1 A definition and a fundamental concept of item response theory and item response modeling

Item response theory (IRT) is an applied statistical and measurement discipline that is concerned with probabilistic functions describing i) the interaction of studied persons and the elements of a measuring instrument or item set of concern, such as items, questions, tasks, testlets, subtests, and subscales (generically referred to henceforth as "items"); and ii) the information contained in the data, which are obtained using the instrument, with respect to its items and overall functioning as well as the examined persons (Reckase 2009).

A fundamental concept in IRT is the relationship between i) the trait, construct, ability, or latent dimension (continuum) being evaluated with the instrument, the dimension being typically denoted θ and often assumed unidimensional but in general may consist of two or more components (see below); and ii) the probability of "correct" response on a given item for a random subject with a trait or an ability level, θ, that is designated as $P(\theta)$.[1] A function of θ, which describes this probability $P(\theta)$ for an item, is called an item characteristic curve (ICC). (Throughout this chapter, we assume that θ is unidimensional unless otherwise indicated.) Owing to its special relevance to IRT and item response modeling (IRM), the ICC can be viewed as one of its main concepts. Other frequent references to it are item response curve, item characteristic function, item response function, or item trace curve. It is important to stress that while being defined as a probability, the ICC is not assumed to be a "static" concept but is rather a function of the (presumed) underlying latent dimension θ. This functional relationship between the probability of a particular response ("correct" response) on a given item and the studied trait, construct, or ability, θ, can be viewed as a characteristic

---

1. As is common in the IRT literature, the notation θ is used throughout this book to denote i) the studied latent trait, ability, construct, continuum, or, in general, latent dimension (or dimensions); ii) an individual value or point on the last (for example, a subject's latent trait or construct score or ability level that is of interest to evaluate); and iii) the horizontal axis of figures of item characteristic curves (ICCs) that imply the consideration or assumption of θ as a single latent continuum or "scale", which is at times also referred to as "θ-scale". While this might be viewed potentially as mixing or even as overusing the symbol θ, strictly speaking, we follow this standard notation that has been widely used in the literature over the past several decades. The specific meaning of its usage, in a sense mentioned in points i) through iii) in this note, is determined by context in the pertinent discussions in the following chapters.

element of IRT and IRM. A prototypical ICC, for the case of a single latent dimension or continuum θ that underlies subject performance on a given set of items or measuring instrument (and a binary scored item), is presented in figure 1.1, which also emphasizes the extended S-shape of this curve.

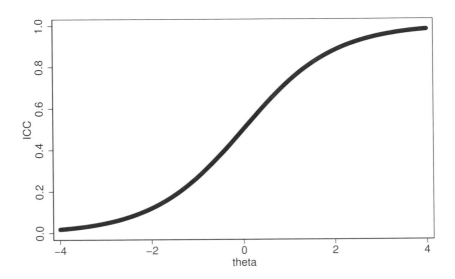

Figure 1.1. A prototypical item characteristic curve

From figure 1.1, we can see that as one moves from left to right on the horizontal axis representing the studied latent dimension θ (referred to as "theta" on the figure), the ICC is fairly low initially, then enters a region of notable increase in the central part of the curve; toward the right end of the presented range of θ, the curve "stabilizes" at a fairly high level, approaching 1 (see chapter 2 for further discussion on the ICC concept). As will be explicated later, different items in a given measuring instrument or item set of concern differ in general in the steepness of the curve increase in that central part of the ICC. The extent of this increase, or curve tilt, is captured by a particular parameter of interest in IRT and its applications in instrument construction and development, which is referred to as item discrimination parameter. (In chapter 5, we discuss in detail this and other item parameters.) Although the majority of applications of IRT are currently based on models that assume the same functional class for the ICCs of all items involved in a studied set or instrument (for example, when all items are binary or binary scored), one can also use IRT with so-called hybrid models. These models allow subsets of items to follow different functional classes for their ICCs, for example, when some items are binary whereas others are ordinal and with more than two available response options (see chapter 11).

While individual items are a special focus of IRT and IRM, measuring instruments consisting of multiple items are also of particular interest. Such instruments—for instance scales, tests, test batteries, surveys, questionnaires, self-reports, inventories, subscales, or testlets—are highly popular in the behavioral, educational, and social sciences (for example, Raykov and Marcoulides [2011]). Their popularity in these and cognate disciplines is to a large degree due to their being composed of multiple components, which provide converging pieces of information about underlying traits, abilities, and attitudes, or in general latent dimensions that are often referred to as "constructs". These constructs and their relationships with one another and with other variables are of main interest in those and related sciences. The reason is that entire theories in them are advanced and developed in terms of such indirectly observable, latent, or hidden variables. This is because the latter typically reflect substantively important theoretical concepts of special concern in these and cognate disciplines. The latent variables can be defined as random variables that presumably possess individual realizations in all subjects in a studied population (or a sample from it), while no observations are available on their realizations (for example, Bollen [1989]). These variables are unobserved, however, because they cannot be directly measured, assessed, or evaluated. They are assumed to be continuous throughout this book, and information about them is collected in their manifestations, proxies, or indicators in observed behavior. (See, for instance, Raykov, Marcoulides, and Chang [2016] and references therein for alternative settings with discrete latent variables.) As such latent variable manifestations, one can usually consider the responses obtained from the studied subjects on the components or elements of instruments used to evaluate the unobserved constructs. Thereby, in the role of instrument components, one typically uses appropriate items, such as questions, tasks, or problems to solve (for example, McDonald [1999]).

The present book, as alluded to earlier, deals with a particular approach to the study of the interaction of persons (respondents, examinees, patients, etc.) with measuring instruments and especially with their elements or components (items). The aim thereby is to optimally use the information about persons and items that is contained in the subject responses to the items. We will be specifically concerned with these responses on the items as well as the studied persons' performance on the considered instruments.

As indicated above, a major focus in IRT is on the relationship between i) the probability P of a particular type of response (such as "correct", "true", "present", endorsed, "success", "agreed", "applicable", etc.) on any given item; and ii) the underlying presumed latent dimension (or dimensions) of interest to evaluate, such as ability, proficiency, trait, construct, or attitude (latent continuum), typically denoted θ in IRT contexts (see also footnote 1). That is, throughout this book, we will be especially interested in the above function $P(\theta)$ describing this relationship—called ICC as mentioned earlier—for each of the items in a measuring instrument or item set under consideration. With this in mind, one could define IRT in simple terms as a methodology dealing with modeling the function $P(\theta)$, that is, expressing in quantitative terms the relationship between θ and the above probability as a function of θ. This probability function includes specific characteristics (parameters) of the used items and of the studied subjects, with the items usually representing a measuring instrument of interest. In fact, most contemporary IRT applications can be seen as essentially concerned with the following activities (no ranking is implied in terms of their relevance):

   a) postulating models about this relationship, that is, for $P(\theta)$, that involve unknown parameters associated with the items of the instrument;

   b) estimating these parameters using an available dataset obtained with the instrument;

   c) evaluating the (relative) fit of the models used; and,

   d) based on the results of the activities in a) through c), estimating (predicting) individual subject values for θ using plausible (selected, preferred) models, with the values being positioned on the same "scale" or underlying latent dimension or continuum as are particular item parameters (when θ is unidimensional; see below).

An especially important and useful feature of IRT is that at the end of its application, using a plausible model for an available dataset from a given item set or instrument one obtains the following two sets of quantities that are commensurate, that is, located on the same continuum (when unidimensional) (for example, van der Linden [2016b]):

   i) a set of (estimated) quantities or parameters characterizing the items, specifically, their difficulty parameters (see chapter 5 for a more precise definition); and

   ii) a set of quantities or values (predictions, assigned values, or estimates—one per person in unidimensional IRT and more than one in multidimensional IRT) that characterize the extent to which each person possesses the trait or traits being evaluated with the instrument in question (see chapters 5, 6, and 12 for further details and examples).

Based on this discussion, we can observe the following important fact. For a given person and binary or binary scored item (measure), the function $P = P(\theta)$ is actually the mean of his or her response or observed or recorded score on the item. This observed score is in general a random variable, and we will denote it by Y throughout the book, also when it is nominal or ordinal with more than two possible responses (see Raykov and Marcoulides [2013] and also chapters 3 and 11). We emphasize that it is the latent dimension (or dimensions), θ, that is of actual interest to measure. However, as we indicated earlier, this is not possible to achieve in any way similar to how one can measure, say, length or weight. In particular, there is no "ruler" or (weight) "scale" that is available to accomplish this measurement. Instead, only the crude assessment of θ is feasible, namely, by using the above indicated measurement process using the presumed proxies, indicators, or manifestations of θ, that is, the items or instrument components. This process consists of administering the set of items or instrument used and recording subject responses to them. The process produces as a result the observed (recorded, manifest) Y scores of the studied persons on the individual items. This complex set of what are actually indirect measurement activities with respect to θ that yield in the end the observed scores on the items is followed in IRM by suitable modeling and estimation procedures that are of main concern in the remainder of the book.

As far as the ICCs are concerned, we should stress that there are infinitely many possible choices of the function $P(\theta)$ for any given item (instrument component). However, as seen later in this book, only a few have obtained prominence and are used most of the time in current behavioral, educational, and social research. The specifics of these choices are attended to in later chapters.

A simple representation of the aims of IRT and IRM can be found in the following schematic, figure 1.2. This figure is used only for conceptual purposes here and is not meant to be a model or "causal" path diagram; that is, no causal implications are intended to be drawn from it ($k > 1$ is the number of items in a set, test, scale, or, more generally, measuring instrument of concern). The graphic representation in figure 1.2 makes use for the current aims of what has been often referred to as path diagram notation (for example, Raykov and Marcoulides [2006; 2011]). In this widely used notation, i) rectangles denote observed variables, or items (that is, the variables that we collect or record data on), while ii) ellipses symbolize the unobservable (latent) variable or variables that we are interested in making inferences about based on the obtained data on the items. As pointed out above, it is the latent variable, trait, or construct—in general, latent dimensions or continua—that the observed measures presumably contain information about. Hence, it is of interest to "extract" this information via appropriate use of optimal statistical methods, like those offered by IRT.

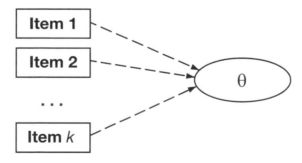

Figure 1.2. A conceptual representation of the aims of item response theory and item response modeling

In figure 1.2, on the left is what we observe, that is, have or collect or record data on, namely, the k items usually representing a measuring instrument of concern. On the right of this figure is what we want to make inferences about using those data—namely, the latent trait, construct, or ability, θ. The connections between the items and θ, represented informally by the one-way arrows in the middle of the figure, are facilitated by the assumed ICC for each item when θ is unidimensional (see chapter 12 for the multidimensional case). These ICCs are typically taken to be monotonically increasing

```
. twoway function fi = normalden(x), range(-4 4)
```

This command produces the standard normal PDF graph displayed in figure 2.1.

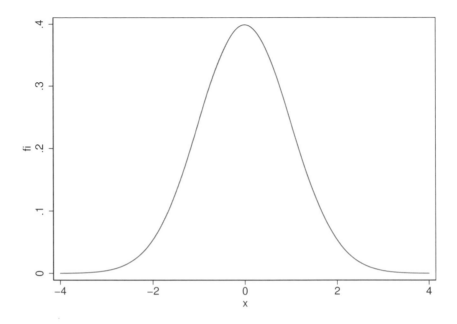

Figure 2.1.   Graph of the probability density function (density curve), $\varphi(x)$, of the standard normal distribution $N(0,1)$ [see (2.2)]

As we alluded to earlier, there is another interesting, graph-related interpretation of the standard deviation, σ, of any normal distribution. Specifically, it can be shown using calculus that the distance from the mean, in either direction, to the so-called inflection point of the curve $f(x)$ in the above (2.1) is precisely σ (for example, Zorich [2016]). Each normal distribution PDF has two such points, a right inflection point above the mean and a left inflection point below the mean. The notion of inflection point will be of special relevance later in the book when we again discuss ICCs in the context of particular IRT models (for example, chapter 5; see also chapter 1). This point for a normal distribution, say, the right inflection point (and by analogy for the left inflection point), is located at that position on the horizontal axis where the curvature of the PDF $f(x)$ in (2.1) changes from concave to convex. That is, the right inflection point is at that location on the horizontal axis for the density curve where the tangent to the PDF "moves" from being above that curve to being below it as one lets the argument x in (2.1) increase, starting with a small positive number (say, when the mean is not negative, that is, $\mu \geq 0$). In particular, in figure 2.1 representing the standard normal PDF, the right inflection point can be shown using calculus to be at 1. That is, this point is actually at σ because the latter is 1 there. Similarly, the left inflection point for this PDF is at -1, that is, at $(-\sigma)$.

Looking again at figure 2.1, one may also wonder why we restricted our attention to the interval $(-4, 4)$ for the normal distribution of concern in it. The reason is that the probability of picking at random scores following the standard normal distribution, which are lower than -4 or larger than 4, is essentially 0 (for example, Agresti and Finlay [2009]). This fact will also be important for us later in the chapter and book, where we use Stata for graphical presentation purposes. More specifically, we will need to first find out, or know beforehand, what particular area or domain of values for the argument (say, x as above) of the functions of interest are of relevance to base the desired graph on them.

2.1.2 The normal ogive function

The standard normal distribution, $N(0, 1)$, and particularly the fact that it has a mean of 0 and standard deviation of 1, leads us to an important concept in IRT that has been traditionally used across several scientific disciplines over the past century or so, as mentioned in chapter 1. This is the notion of the normal ogive curve (or normal ogive for short). As indicated earlier, the normal ogive is defined as the CDF of the standard normal distribution, that is, the curve of the standard normal probabilities (for example, Raykov and Marcoulides [2013]). This curve is often denoted $\Phi(x)$ and is defined for all $x(-\infty < x < \infty)$. (As we mentioned before, while it can be said that in general there are infinitely many normal ogive curves—the CDFs of all possible, infinitely many normal distributions—in this chapter and the rest of the book, we will use the reference "normal ogive" only for the CDF of the standard normal distribution.)

The normal ogive $\Phi(x)$ is of particular interest in this chapter, and so we want to take a closer look at it. We graph it with Stata as follows (see figure 2.2 displayed after the following command, where `Fi` stands for the Greek letter Φ):

An even simpler way to graph the logistic function is actually by using the following Stata command, which yields an identical in all respects plot to that already displayed in figure 2.4 (not presented subsequently, for this reason):

```
. twoway function Lambda = logistic(x), range(-4 4)
  (output omitted)
```

As we noticed from figure 2.4 and mentioned earlier, the shape of the logistic function is quite similar to that of the normal ogive. This will turn out to be a very helpful fact in the remainder of the book, which we will substantially benefit from. In particular, we emphasize that just like the normal ogive, in the central part of the graph in figure 2.4, the logistic increases most rapidly, in contrast with its behavior outside it (in that figure). In addition, the continuity of the logistic is similarly observed by noting that its graph could be drawn "without lifting the pen". The smoothness of the logistic is observed by realizing that there are no sharp changes (edges, wiggles, or kinks) in its graph. In part due to these marked similarities between the logistic function and the normal ogive, the former may be referred to as the logistic ogive (logistic ogive curve, logistic curve, or standard logistic ogive), as we will do at times in the remainder of the book.

In this context of comparing the normal ogive with the logistic function (logistic ogive), it is worthwhile also pointing out a consequential difference between them. Indeed, from the preceding discussion in this section, the following observation is readily made. While the logistic function $\Lambda(x)$ equals the integral in the right-hand side of (2.7), evaluating $\Lambda(x)$ for any real number x does not actually need to involve this integration. In fact, as we see from (2.5), the corresponding value of $\Lambda(x)$ is directly obtained by dividing the numerator to the denominator in its right-hand side. This is a (relatively) straightforward activity that could even be performed with a hand-held calculator. In particular, this will not include the far more involved numerical process of integration, which as mentioned earlier is needed for evaluation of the normal ogive $\Phi(x)$. This practically consequential tractability difference is the main reason why the logistic IRT models, which are based on the logistic function and discussed in detail in the subsequent chapters of the book, have received far more interest and attention by methodologists and substantive researchers over the past several decades than counterpart normal ogive IRT models based on the normal ogive.

We discuss next another feature shared between the normal ogive and logistic curve, that of invertibility. In addition, we will be concerned with the related concepts of odds and logit, which will be of special relevance for the remaining chapters.

2.2.2   Invertibility of the logistic function, odds, and logits

From figure 2.4, we readily make an important observation, which we have in effect already alluded to in the preceding subsection. Specifically, because the logistic is a monotonically increasing function as mentioned above, it is also invertible. That is, for each given point on the vertical axis in figure 2.4, there is a unique number x on its horizontal axis with the property that $\Lambda(x)$ equals precisely that initial point (value). In fact, to obtain the inverse function to the logistic function, all one needs to do is solve its defining equation (2.5) in terms of x, after denoting as p, say, the ratio on the right-hand side of (2.5). [See the earlier discussion in this chapter, and note that this ratio p resides within the interval $(0,1)$.] This leads, after some direct algebraic rearrangements, to the following result,

$$x = \ln\{p/(1-p)\} = x(p) \tag{2.8}$$

where $\ln(\cdot)$ denotes natural logarithm as in the rest of this book. (We recall here that this is the logarithm with basis the Euler constant $e = 2.718\ldots$). In the last part of (2.8), by stating the equality to $x(p)$, we emphasize that x is actually a function of p, namely, the inverse function of the logistic function $\Lambda(x)$. [This inverse is in fact the logit function, as stated in the middle of (2.8); see next discussion.]

Equation (2.8) defines what is referred to as the logit function (or just "logit") and can be formally reexpressed as

$$x = \text{logit}(p)$$

where $\text{logit}(\cdot)$ denotes this function. We notice readily from (2.8) that the logit function is defined for all values of p between 0 and 1, that is, for all p from the interval $(0,1)$. (These values of p will be of typical interest in this book, and we provide a graph of the logit function shortly, after discussing the ratio $p/(1-p)$ that it is based on.) Relatedly, the point at which the logarithm in the middle part of (2.8) is taken, namely, the ratio

$$w = p/(1-p) \tag{2.9}$$

is called "odds". We notice that the odds are defined for all probabilities p from the interval $[0,1)$. The reason for their name "odds" is that w represents the odds for the event "correct solution" (say, in a behavioral or educational testing context) to occur, when the probability of this event is p. More precisely, these are the odds of the event "a value of 1 is obtained on a binary random variable of interest", such as the answer on a binary or binary scored item or question scored 1 versus 0. With this reference in mind, the logit as defined in (2.8) is also called log odds. (This is actually the name under which the middle part of (2.8), that is, the logit, is widely known in the medical, biomedical, and bio-behavioral disciplines.)

We can readily obtain with Stata a graph of the odds w using the following command (note the software requirement to use the argument x in the function, rather than another symbol):

```
. twoway function odds = x/(1-x), range(0 1)
```

This furnishes the graph in figure 2.5 (the notation on the horizontal axis represents formally the probability p that the odds w are a function of).

In (2.13), the first appearance of "x" is on the modified scale mentioned above that results from "compressing" the initial units of measurement (scale) by a factor of 1.701, which renders x^* from x. We use this useful fact next.

2.3.1 Expressing event or response probability in two distinct ways

The discussion in section 2.3 in effect showed that we have a choice of two functions to express the probability of an event or given response to a binary or binary scored item under consideration in relation to other quantities of possible interest. Specifically, we can choose between i) the inverse logistic (that is, the logit, or logit function) and ii) the inverse normal ogive function, which is referred to as the probit (or probit function) as mentioned earlier. In other words, either of these two functions, logit or probit, represents formally the relationship between i) a probability of an event or a particular response to a stimulus, such as an item with two possible responses or scores; and ii) an underlying number, such as x above (after a rescaling if need be, as indicated earlier).

In the majority of the following chapters, it will be of relevance to be in a position to deal with the probability of the event "correct response" to a given item (question, problem, task). We will assume that this response is denoted (scored) "1" in the remainder of the book. (In general, this can be any prespecified response option on a binary or binary scored item.) In particular, we will be concerned with expressing this probability as a function of other more fundamental and substantively interpretable quantities, typically referred to as parameters, which are generally not limited in magnitude. The preceding discussion in this chapter has actually given us two essentially equivalent ways to achieve this goal for the aims of our developments in the book. Specifically, we can model this probability using either the normal ogive or the logistic function.

As pointed out earlier in this section, however, primarily for mathematical and numerical convenience and especially tractability, the logit function turns out to be easier to work with for most modeling, analytic, and empirical purposes in relation to IRT and IRM (compare Cai and Thissen [2015]). Thus, we will adopt the use of the logistic function in the remainder of the book. At the same time, we note that the probit function has also a number of important uses and in fact is no less applicable for the same purposes. However, as indicated previously, the probit function is more tedious to work with and less numerically tractable than the logistic function; hence, it will not be our choice when it comes to using particular IRT models in the rest of the book.

2.3.2 Alternative response probability as closely related to the logistic function

We mentioned in section 2.2 that we can consider the logistic function $\Lambda(x)$ as producing, for any given number x, the probability of an event ($-\infty < x < \infty$). Hence, the probability of its complementary event is furnished by the function $\Lambda_0(x) = 1 - \Lambda(x)$. Suppose we consider now the event of interest being "correct" response (that is, denoted "1") on a binary or binary scored item. The probability of this event is thus representable by the logistic function $\Lambda(x)$. It follows then from the preceding discussion that the probability of the alternative response on the same item, namely, "incorrect" (assumed to be denoted "0" in the remainder of the book), is representable as $\Lambda_0(x)$ (compare Baker and Kim [2004]).

We can readily graph this function $\Lambda_0(x)$ using Stata as follows [the plot is presented after the command; Lambda_0 stands for $\Lambda_0(x)$]:

```
. twoway function Lambda_0 = 1-logistic(x), range(-4 4)
```

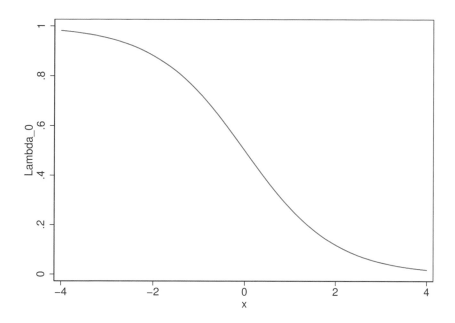

Figure 2.8. Graph of the function $\Lambda_0(x)$

We easily notice from figure 2.8 that the shape of the function, $\Lambda_0(x)$, is in a sense the "reverse" of that of the logistic function, $\Lambda(x)$ (compare figure 2.4). In particular, $\Lambda_0(x)$ initially decreases slowly and in the "central part" of the presented range on the horizontal axis accelerates this negative trend; toward its right end, its decrease becomes quite gradual again. In fact, the graph of $\Lambda_0(x)$ is exactly symmetric to that of $\Lambda(x)$ (over the same range of x), with the symmetry axis being, for instance, the vertical line erected at the point $x = 0$ on figures 2.4 and 2.8. (See below for further details about the intersection point of $\Lambda_0(x)$ with $\Lambda(x)$, which is actually the inflection point of $\Lambda(x)$ as well; see also chapter 1.)

We can readily observe these features of $\Lambda_0(x)$ by overlaying the graphs of both functions, $\Lambda_0(x)$ and $\Lambda(x)$.

```
. twoway function Lambda = logistic(x), range(-4 4)||
> function Logistic_0 = 1 - logistic(x), range(-4 4)
```

This command furnishes the following graph.

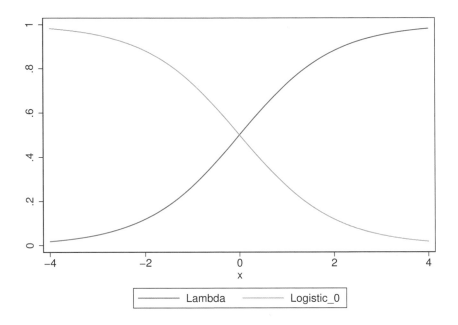

Figure 2.9. Graph of the functions $\Lambda_0(x)$ and $\Lambda(x)$

In figure 2.9, the previously mentioned fact that the sum of the functions $\Lambda_0(x)$ and $\Lambda(x)$ equals 1 is easily noted for each number x. Indeed, at any point on the horizontal axis, adding the pertinent values of these functions obviously yields 1 (see also these functions' definitional equations in section 2.2). In addition, as one approaches and passes the point $x = 0$ on this axis (with smaller values), the probability represented by $\Lambda_0(x)$ behaves as follows: i) first, it stays higher than the probability expressed by $\Lambda(x)$; ii) then, it equals the latter at the zero point only; and iii) finally, immediately past that point, the relationship of these functions reverses direction, with $\Lambda(x)$ being higher as a probability than that given by $\Lambda_0(x)$. That is, before the point $x = 0$, the answer denoted 0 ("incorrect") is more probable; at $x = 0$, it is as probable as the answer denoted 1 ("correct"); and after $x = 0$, the latter response is more probable than that denoted 0. Hence, when presented with an item whose item ICC is $\Lambda(x)$ (see figure 2.9), a person with a trait or an ability level θ that is lower than that represented by the point $x = 0$ (that is, with $\theta < 0$ in our earlier, alternative notation) is more likely to respond 0, or incorrectly, on it. However, at $x = 0$, that is, if possessing ability level represented by the point $\theta = 0$, he or she is as likely to respond 0 (incorrectly) as he or she is to answer 1, or correctly. After $x = 0$, that is, if he or she has an ability level above the one represented by the point 0 (that is, with $\theta > 0$), that subject is more likely to respond 1 on the item of concern.

Further, from figure 2.9, we observe that the two graphs of the functions $\Lambda_0(x)$ and $\Lambda(x)$ are in fact symmetric to each other in two aspects. Specifically, they are i) symmetric with respect to the so-called mid-probability line that is drawn horizontally at the level of 0.5 probability (on the vertical axis); and ii) symmetric with respect to the vertical line drawn at $x = 0$, as mentioned before. Moreover, the tangents to either of these two functions at that point are also symmetric images of each other with respect to the same vertical line at $x = 0$. In addition, these tangents intersect with the horizontal axis at angles with the same magnitude but "open" in opposite directions. More concretely, for $\Lambda_0(x)$, that angle is "facing" the negative numbers, and for $\Lambda(x)$, the positive numbers.

Furthermore, the intersection point of the two functions of concern, $\Lambda_0(x)$ and $\Lambda(x)$, represents an inflection point for either of them. (This can be shown using calculus, noting that their second derivatives vanish there; for example, see Zorich [2016].) Specifically, at that inflection point, the PDF of the logistic distribution attains its maximum [see (2.6), which presents the derivative of the CDF of the logistic distribution in (2.7); see also figure 2.3.] We also notice that before this point, the curve of $\Lambda(x)$ is convex, whereas it is concave after it; the converse is true for the function $\Lambda_0(x)$. Last but not least, if the ICC on a given binary item can be well represented by a logistic function (see chapter 1 and figure 2.4), then the decreasing function $\Lambda_0(x)$ can be seen as the ICC for the same item if it is reverse coded. That is, $\Lambda_0(x)$ would be the original item's ICC if the latter were to be defined as the probability of "incorrect" response on that item to begin with (as a function of x, or θ in our earlier alternative notation).

These readily made observations from the last four figures and preceding discussion in this chapter will turn out to be quite useful in the developments provided in chapter 11. As indicated earlier, we will be considering there polytomous items and associated IRT models. These items generalize the currently considered case of binary or binary scored items to that with more than two possible response options.

2.4   Chapter conclusion

In this chapter, we have discussed two main functions underlying applications of IRT and IRM in the behavioral, educational, and social sciences as well as related disciplines— the logistic function and the normal ogive. They are of basic relevance for the so-called logistic and normal ogive IRT models in this measurement field, respectively. We will use the former models in the rest of the book because of mathematical and numerical tractability and convenience as mentioned in the chapter. We have also familiarized ourselves with the software underlying this book, Stata, which we have used here mainly for graphing of these and related functions. In the chapters to follow, we will discuss further concepts of IRT and their relationships as well as provide numerous empirical applications of this measurement approach using Stata.

been widely circulated in the older literature but also appear in more recent writings.) These misconceptions have been spread for several decades now and effectively inhibit the realization of the important CTT-IRT relationships, which are of relevance also for the remainder of this book (compare Raykov and Marcoulides [2011, 2016b]; see also Kamata and Bauer [2008] and Kohli, Koran, and Henn [2015]).

An incorrect interpretation of CTT found in the literature is that (3.6) represents an assumption of CTT. As indicated above, such a statement is not correct because the uncorrelatedness of true and error score (for a given subject and observed measure or score) already follows from the very definition of true score. Hence, there is no need to advance or refer to (3.6) as an assumption of CTT, because the true with error score uncorrelatedness in (3.6) is always the case, cannot be a false statement, and therefore cannot be an assumption of CTT.

According to another misconception about CTT, the latter assumed the variance of the error term E for a given observed measure (score) Y as constant across all subjects. This is incorrect as well. Specifically, such an assumption is never made in the derivation of the true score, and thus of the error score, as one can also see from the preceding discussion in this section for the definition of the true score (see Zimmerman [1975] and also footnote 1). Even if one focuses only on one observed measure (score), Y, the obvious implication is that it is considered an RV at the subject population level (see section 3.1). Hence, the studied subjects' values on it are its individual realizations (for example, in a sample from that population). For this measure, therefore, the concept of subject level error score (as part of the RV Y) is not defined, applicable, or relevant. Hence, this concept is meaningless then. Therefore, no statement about it would make sense. Relatedly, if one considers in the same setting the notion of reliability, for the same reasons, one does not need to assume identity of error variance across subjects in order for this notion to be applicable and meaningful. This is because the reliability coefficient, as a ratio of true to observed variance, obviously is meaningful at the considered population level without any further assumptions, of course as long as the observed variance is not 0, as it is assumed to be the case for any manifest measure throughout this book (compare Raykov and Marcoulides [2011]; see also chapter 2).[2]

A third misconception about CTT is that it is based on a model, referred to as the "CTT model". Circulators of this misconception point to the above CTT decomposition in (3.1) as that model. This is an incorrect treatment, interpretation, and reference to the CTT decomposition in (3.1) [see also (3.2) and (3.3)]. The reason is that from a logical (and philosophy of science) standpoint, a model can be only a proposition based on a set of assumptions. The latter are by definition statements that can at least in principle be falsified in an empirical setting, that is, could be incorrect. However, the CTT decomposition in (3.1) [see also (3.3)] is always correct. This is because it effectively states the decomposition of any given observed measure or score (RV) into its

---

2. If one alternatively considers the score Y_{ji} for a given measure and subject [see (3.3)], the existence of its true score T_{ji} is ensured always, that is, with no need for any additional assumptions with respect to the error score E_{ji} in (3.5), such as say error variance constancy across persons, $i = 1, \ldots, n, j = 1, \ldots, k$ [the same argument applies if considering instead the score Y_i from (3.2)].

mean plus mean deviation (!). Because this decomposition is always feasible and true, it can never be wrong, that is, can never be falsified. Thus, the CTT decomposition cannot be a model. In fact, there cannot be a CTT model that is based only on this equation for a given observed measure. (See section 3.3 below for models that are empirically indistinguishable from models based on CTT, or CTT-based models, which involve more than a single measure or observed score. Such CTT-based models are the popular congeneric, true score equivalent, and parallel test models and are each falsifiable in general; see, for example, Jöreskog [1971], and Raykov and Marcoulides [2011, 2016b]).[3]

To deal with yet another misconception that is particularly misleading in the context of binary or binary scored items of concern in the majority of the chapters of this book, we believe it is necessary to remind ourselves of a basic statistics relationship that has important consequences for our subsequent discussion of IRT and IRM. We deal with that relationship next.

3.1.3   Binary random variables: Expectation and probability of a pre-specified response

A binary or binary scored item, frequently referred to also as a dichotomous item (measure), is a typical example of a binary RV. Such a variable is at times also referred to as an RV following the Bernoulli distribution, or a Bernoulli RV (for example, Agresti [2013]). More concretely, suppose Y_{ji} is an RV associated with a dichotomous item, that is, represents the observed score on the item, and is thus a binary RV ($1 \leq i \leq n, 1 \leq j \leq k$; see below). Assume then, as is usual, that this item can take on the values 0 and 1. Let us also assume that the value 1 is assigned to a particular prespecified response on the item, such as "true", "correct", "endorse", "present", "applies to me", "agree", "success" or similar (for simplicity referred to as "correct" in the rest of this book), or another response of interest that is fixed beforehand. Then suppose that

$$Y_{ji} = \begin{cases} 1, \text{with probability } p_{ji}, \text{ say, and} \\ 0, \text{with probability } 1 - p_{ji} \end{cases}$$

For this discrete RV, Y_{ij}, its expectation is (for example, Casella and Berger [2002])

$$\mathcal{E}(Y_{ji}) = 1 \times p_{ji} + 0 \times (1 - p_{ji}) = p_{ji} \tag{3.7}$$

---

3. A "comparative" discussion, if deemed meaningful, of what can be achieved using CTT and IRT must obviously treat both of them on an equal footing to begin with (for example, Raykov and Marcoulides [2016b]). For this to be the case, a set of items must first be given and CTT as well as IRT be used on the entire set, rather than the former on a single item only [as in (3.1)] and the latter on the full set as found in multiple "comparative" but misleading discussions of CTT and IRT in the past (as well as in some more recent measurement-related accounts). When that equal footing is ensured with a given set of dichotomous items, the discussion in Raykov and Marcoulides (2016b) shows the equivalence of a correct CTT-based and an IRT-based measurement approach in the setting considered in that source (see also Takane and de Leeuw [1987]; Kohli, Koran, and Henn [2015]).

where $\times$ denotes multiplication. That is, the probability of "correct" response on this item, as represented by the RV Y_{ji}, is identical to its expectation or mean. We stress that this expectation is defined for any value of the above probability for the response denoted 1, namely, p_{ji}, which resides in the closed interval $[0, 1]$ ($1 \leq i \leq n, 1 \leq j \leq k$).

We provided the above discussion in this subsection to deal with a rather consequential misconception about CTT that is surprisingly widely circulated. According to that misconception, CTT was applicable only when the measure in question (the Y score in this chapter) was on an interval scale, that is, only when Y was a continuous RV or could be treated as such. This statement (misconception) is incorrect, however, because all that is needed for CTT to be applicable is the existence of the true score associated with a given observed score (see also footnote 1). Yet the true score existence is ensured in contemporary behavioral, educational, and social research as well as in empirical research more generally, as mentioned before, because this research yields only bounded observed measures and scores or RVs, Y. Because their expectations $\mathcal{E}(Y)$ exist even for binary (or binary scored) items, as explicated in the above (3.7), it obviously follows that CTT is also applicable to a measure or observed score whose scale is very far from a continuous RV scale, namely, to a binary item. (In the same way, it is shown that CTT is also applicable with ordinal measures, such as the popular and widely used Likert-type items; see also footnote 4 below.)

This reasoning is just as applicable for debunking what may be considered another closely related misconception about CTT. That misconception states that CTT could be applied only on the overall sum score (or weighted sum score) for a given multicomponent measuring instrument or item set but not on its individual items.[4] That this is also an incorrect statement, and hence that CTT is just as applicable on the individual instrument components, items, or measures, is readily realized by referring to the rebuttal of the last discussed myth (see preceding paragraph).

3.2   Why classical test theory?

We have emphasized in chapter 1 that a main focus in IRT is on the relationship between studied latent traits and the probability of a particular response on a given item presumably tapping into it, say, of "correct" answer (the response symbolized or scored by 1) for a subject with trait level θ. That is, for a given binary or binary scored item, of main concern in IRT is the relation between θ and the probability $P(\theta)$ for "correct" response on it by such a person.

---

4. We point out that the true score is defined, as above in this chapter, also for a nominal item. Specifically, the true score on such an item is defined as the mean (expectation) of its associated observed score Y, as an RV, on the "scale" defined formally by the numerals used to represent the item categories. This definition is logically sound and meaningful per se, irrespective of the fact that the metric of the pertinent observed variable is uninterpretable in substantive terms for such an item. We observe that the empirical interpretation of the true score cannot be free then from the latter limitations that are inherent in the observed score metric to begin with.

However, as we have just seen in section 3.1.2, the probability of "correct" response is precisely the mean of the observed binary item score, Y_{ji}, for a given item and person, that is, for a given trait level θ $(1 \leq i \leq n, 1 \leq j \leq k)$. Yet this mean is exactly the true score of Y_{ji}, namely, T_{ji}, as can be seen also from (3.4) and (3.7). Hence, for a binary or binary scored item (also referred to at times as a "dichotomous item"), a main focus in IRT is in fact on its own true score, namely, on

$$P_j(\theta_i) = \mathcal{E}(Y_{ji}) = T_{ji} = T_{ji}(\theta_i) \qquad (3.8)$$

where $P_j(\theta_i)$ denotes formally the probability of "correct" response on the jth item by the ith person possessing the trait level θ_i. That is, in the dichotomous item setting considered here, which spans a very large part of contemporary applications of IRT, particularly in the educational and behavioral disciplines, IRT and IRM mainly focus on the CTT concept of true score (as a function of the underlying trait, θ). We also note that $E_{ji} = Y_{ji} - P_j(\theta_i)$ is in fact the CTT error score associated with that item. (Notice also the functional dependence of the error score, E_{ji}, on the individual trait or ability level, θ_i, which fact is obviously incompatible with an assumption of its constancy across subjects, according to an earlier discussed misconception in section 3.1.2; $i = 1, \ldots, n$; see also footnote 2.)

We close this section by noting that from (3.8) follows that the fundamental concept of CTT, the true score, is in the binary or binary scored item case (perfectly) nonlinearly related to the underlying latent trait, construct, or ability being evaluated by it, θ, which is of major concern in IRT. That is, the true score is a nonlinear function of the trait, construct, or ability of focal interest in IRT and IRM. With this insight, we are now ready to move on to another topic that is also closely related to IRT, that of classical (linear) factor analysis (CLFA).

3.3   A short introduction to classical factor analysis

As we showed earlier in the chapter, there is no CTT model per se if considering a single observed measure. However, there are several useful models for empirical behavioral, educational, and social research that can be developed using CTT when more than one observed measure is simultaneously considered (rather than each measure in isolation from the others). These models are referred to as CTT-based models (for example, Raykov and Marcoulides [2011]). Perhaps the most popular of them is the CTM. The CTM assumes the true scores of a given set of k observed (manifest) measures as perfectly linearly related among themselves ($k > 1$; see, for example, Jöreskog [1971]). Special cases of the CTM, which share nearly the same popularity, are i) the true-score equivalent model and ii) the model of parallel tests (for example, McDonald [1999]). In the single measurement occasion setting that underlies this book and arguably the large majority of contemporary routine applications of IRT, the CTM is not empirically distinguishable from another highly popular model. This is the CLFA model that is the subject of the present section. (For potential differentiability between CTT-based and counterpart FA models in the repeated assessment occasion case, see Raykov and Marcoulides [2016a].)

3.3.1   The classical factor analysis model

The model of classical FA "resembles" that of multivariate multiple regression, the general linear model (for example, Timm [2002]). More specifically, the CLFA model relates each observed measure to one or more latent variables, such as traits, abilities, latent dimensions, or continua. To be more concrete, suppose one is interested in k observed interrelated continuous variables or manifest measures, denoted $Y_1, Y_2, \ldots, Y_k (k > 1)$. Let us assume also that there may be m $(0 < m < k)$ more fundamental latent variables, $f_1, f_2, \ldots, f_m$, which are typically referred to as "factors" in an FA context. We posit that these factors may be "responsible" for the observed interrelationships among those measures, for example, for their correlations (see Raykov and Marcoulides [2008]).

The CLFA model consists then of as many equations as there are observed variables and is represented as follows:

$$Y_1 = \mu_1 + a_{11}f_1 + a_{12}f_2 + \cdots + a_{1m}f_m + u_1$$
$$Y_2 = \mu_2 + a_{21}f_1 + a_{22}f_2 + \cdots + a_{2m}f_m + u_2$$
$$\vdots$$
$$Y_k = \mu_k + a_{k1}f_1 + a_{k2}f_2 + \cdots + a_{km}f_m + u_k \tag{3.9}$$

In (3.9), the u's are zero-mean residuals assumed uncorrelated among themselves and with the factors. The latter, like any latent variable, are presumed throughout this book to possess 0 mean and unit variance (to contribute to model identification when fitting the model to data), unless otherwise indicated. In addition, the a's are the weights of the manifest measure on the corresponding factors, which are referred to as factor loadings. These loadings reflect correspondingly the relationships between the respective observed variables and factors. Further, in (3.9), the μ's are measure intercepts, that is, the respective observed variable means. Because all manifest measures Y_1 through Y_k share the same m factors, the latter are called common factors. At the same time, the u's are referred to as unique factors. This reference is motivated by their feature of containing "pure" measurement error as well as all sources of variance in the associated observed measure that are not shared—via the factors—with the remaining $k - 1$ manifest variables (see Mulaik [2009] and also Raykov and Marcoulides [2011]). With this in mind, the common factors can be viewed as the sources of all shared (common) variability among the k observed variables $Y_1, Y_2, \ldots, Y_k$.

We stress that like standard (multivariate) multiple regression, the CLFA model is a linear model, as is directly seen from (3.9). Indeed, this property of the CLFA model is realized by observing from their right-hand sides that the factors are only i) multiplied by constants—the a's that do not change value across subjects—and then ii) added up and to the unique factors, rather than transformed nonlinearly. With i) and ii), one obtains the observed variables in the left-hand side of (3.9), where these variables appear in their "raw metric" (that is, are left untransformed there). Similarly, we notice that iii) none of the a's are themselves subjected to any transformation before being multiplied with the factors. Based on the features i) through iii), the CLFA model (3.9) is readily seen as a linear model. In contrast to (3.9), in the next section and chapter, we will be discussing nonlinear FA models. While they are conceptually fairly similar to the CLFA model (3.9), they possess the essential feature that unlike the latter, they are nonlinear models. Their important similarity to CLFA models, however, is part of the reason why we have discussed FA in this section.

3.3.2 Model parameters

The informal analogy to regression analysis (general linear model) mentioned above in this section is also useful in providing a relatively simple way in which we can find out the parameters of a CLFA model. (We stress that while helpful, this analogy is not meant to imply treatment of FA models like standard regression models. The reason is that the "predictors" in the former models are not observable; hence, no observations on them exist or are available, unlike the case with those regression models.) This parameter discussion will be of particular relevance when we start working later in the book with popular IRT models, which are conceptual relatives of CLFA (FA) models (compare Raykov and Marcoulides [2011]). Specifically, with this analogy in mind and based on the preceding discussion on the CLFA model, one determines that its parameters are

1. the factor loadings, as counterparts of the (partial) regression coefficients whose role they play in (3.9) (abstracting for a moment from the fact that the factors are not observed, unlike the independent or explanatory variables in regression analysis);

2. the variances of the unique factors, as counterparts of the model error variances (standard error of estimate in the case of univariate simple or multiple regression), as well as the following unique to CLFA parameters;

3. factor correlations (covariances).

We also mention in passing that the CLFA model defined in (3.9) can be more compactly represented using matrices and vectors as

$$\mathbf{Y} = \boldsymbol{\mu} + \mathbf{A}\mathbf{f} + \mathbf{u} \qquad (3.10)$$

In (3.10), $\mathbf{A} = [a_{jr}]$ is the matrix of factor loadings, $\mathbf{Y} = (Y_1, Y_2, \ldots, Y_k)'$ is the vector (set) of observed variables, $\mathbf{f} = (f_1, f_2, \ldots, f_m)'$ is that of the common factors, $\boldsymbol{\mu} = (\mu_1, \mu_2, \ldots, \mu_k)'$ is the vector of observed means (intercepts), and $\mathbf{u} = (u_1, u_2, \ldots, u_k)'$ is the vector of unique factors (priming is used in the rest of the book to denote transposition, unless otherwise indicated; see, for example, Mulaik [2009]).

From (3.10), using familiar rules for working out variances and covariances of linear combinations of RVs (for example, Raykov and Marcoulides [2006]), one obtains in compact matrix terms

$$\Sigma_{YY} = \mathbf{A}\mathbf{\Phi}\mathbf{A}' + \mathbf{\Theta} \tag{3.11}$$

In (3.11), which is a rather popular implication from the model in (3.9), Σ_{YY} is the covariance matrix of the observed variables, $\mathbf{\Phi}$ that of the latent factors, and $\mathbf{\Theta}$ is the covariance matrix of the unique factors (compare Raykov and Marcoulides [2008]). Equation (3.11) succinctly provides the CLFA model consequences with respect to observed variances and covariances. Further, (3.11) lets us also easily locate (most of) the CLFA model parameters (assuming model identification). Indeed, looking at its right-hand side, these are found as i) the elements of the factor-loading matrix $\mathbf{A}$, ii) the elements of the covariance matrix $\mathbf{\Phi}$ of the factors, and iii) the main diagonal elements of the covariance matrix $\mathbf{\Theta}$ of the unique factor vector $\mathbf{u}$ (with $\mathbf{\Theta}$ assumed diagonal, as mentioned earlier). We note that while the observed variable means in the vector $\boldsymbol{\mu}$ are also model parameters, they do not feature in (3.11). This is because the latter equation is an implication of (3.10) with respect to the observed variable covariance matrix only.

3.3.3 Classical factor analysis and measure correlation for fixed factor values

If we take a look at (3.9) just for a couple of manifest measures, say, Y_1 and Y_2, we will observe a rather interesting fact that connects CLFA with IRT. To this end, let us first restate these equations for convenience:

$$Y_1 = \mu_1 + a_{11}f_1 + a_{12}f_2 + \cdots + a_{1m}f_m + u_1$$
$$Y_2 = \mu_2 + a_{21}f_1 + a_{22}f_2 + \cdots + a_{2m}f_m + u_2 \tag{3.12}$$

Suppose we now fix the values of the m factors involved in (3.12). That is, assume we take such a subpopulation ("slice") of the studied population, in which all subjects have the same values on these latent constructs, say, $c_1, \ldots, c_m$, respectively, using here the notation c for constant, that is, a quantity with the same value across all persons (we note that one need not assume these c's as equal among themselves). With these fixed factor values, (3.12) yields the following pair of equalities:

$$Y_1 = \mu_1 + a_{11}c_1 + a_{12}c_2 + \cdots + a_{1m}c_m + u_1$$
$$Y_2 = \mu_2 + a_{21}c_1 + a_{22}c_2 + \cdots + a_{2m}c_m + u_2$$

Hence, for their covariance, the following holds (for example, Raykov and Marcoulides [2006]),

$$\mathrm{Cov}(Y_1, Y_2) = a_{11}a_{21}\mathrm{Cov}(c_1, c_1) + \cdots + a_{1m}a_{2m}\mathrm{Cov}(c_m, c_m) + \mathrm{Cov}(u_1, u_2)$$
$$= 0 \tag{3.13}$$

where $\mathrm{Cov}(.,.)$ denotes covariance of the RVs within parentheses. Equation (3.13) results obviously from the fact that the covariances of the constants are 0, while the last covariance on its right-hand side (of its first line) is 0 by the uncorrelated unique factor assumption made earlier in this section.

Equation (3.13) lets us conclude that under the CLFA model (3.9), the correlation between any two observed measures vanishes at a given set of values for the underlying factors. This is an important feature of the CLFA model, which is often referred to as conditional uncorrelatedness. Notice that with normality of the observed measures, this feature is the same as independence of the observed variables involved for given values of the factors (for example, Casella and Berger [2002]). This property is often referred to as conditional independence or local independence (LI) of the manifest variables, because it is achieved at particular fixed, or "local", values for the factors. More concretely, a pair of bivariate normal measures are conditionally independent once fixing the factors, if the CLFA model holds for a set of observed variables including the pair. Then the underlying factors provide an essentially complete "explanation" of the correlations of the manifest variables. This is because the factor variances and covariances (correlations) are in this case the only source of that observed correlation.

The property of LI is of particular importance also in IRT and may be considered a main assumption within its framework (see chapter 5 for a detailed discussion). As it turns out (for example, Bartholomew, Knott, and Moustaki [2011]), LI is one of the most fundamental and unifying characteristics of several modeling frameworks that involve latent variables. These are latent class analysis, IRT, latent profile analysis, and FA.

Returning to our discussion of CLFA, we see it is LI that allows us to observe, while revisiting this modeling framework, that we have not mentioned anything yet about the number m of factors (other than it being positive and obviously an integer number not exceeding that of the observed measures that are factor analyzed). How could one "determine" that number m? This important and potentially controversial issue may be addressed by adhering to the following guideline based on the LI property: Find the smallest number m, such that given the m factors, the observed k analyzed variables, Y_1 through Y_k, are (conditionally) uncorrelated! Accomplishing this in a given empirical setting permits one to develop a solid argument in favor of an answer to the query about number of factors, m, assuming of course that these factors are interpretable in the subject matter domain of application. We emphasize that this requirement of interpretability of the factors is a main criterion for meaningfulness of an FA conducted on a given set of observed variables (for example, Raykov and Marcoulides [2008]).

The above reasoning in relation to LI is especially relevant in IRT. This is in particular the case when interested in resolving the question of dimensionality of the construct θ, that is, the number of its components. (This is the same as the query of dimensionality of what is at times referred to as the underlying latent space; see, for example, Reckase [2009].) We attend to some aspects of this special question in the initial sections of chapter 5. We observe though that this query is effectively the same as that about the number of traits being evaluated by a considered set of items or measuring instrument. This question represents a main issue that needs to be similarly resolved in applications of multidimensional IRT, which we will be concerned with later in the book (see, for example, Reckase [2016]; see also chapter 12).

3.4   Chapter conclusion

In this chapter, we have discussed CTT and CLFA. These applied statistics and measurement frameworks can be seen as historically preceding IRM, in particular in the behavioral, educational, and social sciences. CTT and FA have enjoyed impressive popularity for much of the past century, in part as the only measurement frameworks widely available at the time. They have also found a great deal of applications in behavioral and social research, as is well documented in the literature (see, for example, Mulaik [2009] and Harman [1976]; see also Raykov and Marcoulides [2008, 2011] and references therein).

An important point we have made in this chapter was the identity of the CTT concept of true score, as a function of the underlying trait or ability studied, to the item characteristic curve in the binary or binary scored item case that represents a widely used setting in contemporary empirical applications of IRT and IRM. This identity provides in particular an important link of IRT to CTT. That link represents an instructive connection, which we will further benefit from in the following chapters (see also Raykov and Marcoulides [2016b]). In addition, we have observed that the notion (assumption) of LI conceptually connects IRT with FA as well as with two other popular frameworks based on modeling in terms of latent variables, namely, latent class and latent profile analysis (for example, Bartholomew, Knott, and Moustaki [2011]). Moreover, LI provides an instrumental and highly useful means when fitting such models to empirical data and addressing the query of dimensionality of the underlying latent structure and studied constructs in empirical research (see also chapters 7 and 12).

In the next chapter, we will extend our discussion of the important and beneficial links of IRT and IRM to other closely tied statistical frameworks, the generalized linear model and, in particular, logistic regression as a special case of it.

4 Generalized linear modeling, logistic regression, nonlinear factor analysis, and their links to item response theory and item response modeling

As we have indicated earlier in the book, a main assumption in item response theory (IRT) is that the responses on the items in a given set or a multicomponent measuring instrument can be accounted for by one or more latent traits, constructs, or abilities. IRT is then specifically concerned with the probability of a particular response (say, "correct" response, or the one scored "1") on each item as a function of the level of these traits and item characteristics. While the majority of current applications of IRT assume that there is a single latent trait behind the responses to the items, much progress has been made in the past 20 years or so in multidimensional IRT. The latter is a complex topic that we will be concerned with in chapter 12. We assume until then that there is only one latent dimension or unobserved continuum, symbolized θ, that is presumed to underlie studied persons' performance on a set of items or an instrument under consideration. That is, all examined subjects are positioned at some locations along this latent dimension that are unknown, and their differences in θ are related to the observed differences in their responses on the individual items.

While in IRT, as in factor analysis (FA), the underlying trait is assumed to be continuous, we typically deal with discrete (for example, dichotomous, categorical, or ordinal) items in IRT and item response modeling (IRM). For instance, these are binary or binary scored items, or alternatively Likert-type items with only a few possible answer options. (See chapter 11 for a nominal response model discussion.) However, the major difference between the IRT models and the classical linear factor analysis (CLFA) model in (3.9) (see chapter 3) is the fact that the former are nonlinear models. Also, a special focus in IRT models is on how subjects with different ability levels respond to each member of a considered set of items. This is the main reason for their high popularity, especially in achievement evaluation contexts, but also in many other settings. At the same time, a key similarity between CLFA and IRT lies in their essential use of continuous latent variables, that is, traits, constructs, abilities, or in general unobserved continua. Therefore, both CLFA and IRT can be seen as parts of the highly comprehensive latent variable modeling methodology (for example, Muthén [2002]). Another main similarity between CLFA and IRT is the local independence property or assumption mentioned in the preceding chapter. Specifically, according to the FA model considered there, for a fixed set of factor values, the observed variables are uncorrelated and hence independent in the case of normality. Similarly in IRT, for fixed values of the latent traits, the responses on the set of items of concern are independent of each other. This local independence property is of special relevance also for other latent variable models used in applied statistics and the behavioral, educational, and social sciences as pointed out earlier in the book, namely, latent class and latent profile (finite mixture) analysis models (see also Lubke and Muthén [2005] and Raykov, Marcoulides, and Chang [2016]).

Given the above considerations, we discuss next the comprehensive applied statistics approach of generalized linear modeling. We will be mainly interested in its special case of logistic regression (LR). The reason is that LR may be viewed as a conceptual framework within which IRT and IRM can be positioned (compare Cai and Thissen [2015]). As we will see, IRT models can be accordingly considered as multivariate logistic models with one or more predictors (covariates, explanatory, or independent variables) that, however, are not observed. (Throughout the rest of the book, the term "predictor" will be used in a descriptive role, that is, as referring to explanatory variables, without any causality-related implications.) This view will bring special benefits later in the book when discussing particular IRT models.

4.1 Generalized linear modeling as a statistical methodology for analysis of relationships between response and explanatory variables

4.1.1 The general linear model and its connection to the classical factor analysis model

The widely used general linear model, in the case of a single continuous dependent variable Y and given predictors $X_1, X_2, \ldots, X_q (q \geq 1)$ is based, as is well known, on the following model equation:

$$Y = \beta_0 + \beta_1 X_1 + \beta_2 X_2 + \cdots + \beta_q X_q + e \qquad (4.1)$$

In (4.1), e denotes the model error term that has mean of 0 and is assumed to be unrelated with the predictors, and $\beta_0, \beta_1, \ldots, \beta_q$ are the intercept and partial regression coefficients, respectively (for example, Timm [2002]). Taking expectation from both sides of (4.1), we obtain another widely used form of the regression analysis model,

$$\mathcal{E}(Y) = \beta_0 + \beta_1 X_1 + \beta_2 X_2 + \cdots + \beta_q X_q \qquad (4.2)$$

for a fixed set of predictor values (compare Raykov and Marcoulides [2011, 2013]). Therefore, the univariate multiple regression analysis model (general linear model with a single outcome measure) represents the expectation of the dependent variable as a linear combination of the predictor values, including an intercept term. In statistical terms, (4.2) is often recast as

$$\mathcal{E}(Y|X_1, \ldots, X_k) = \beta_0 + \beta_1 X_1 + \beta_2 X_2 + \cdots + \beta_q X_q$$

where conditional expectation given the predictors appears in its left-hand side. This is the conditional expectation, $\mathcal{E}(Y|\mathbf{X})$, of the dependent variable that is modeled using the predictors (predictor values) in the vector $\mathbf{X}$. It is this conditional expectation that is assumed to be a linear function of the explanatory variables.

Similarly, an important relationship is readily observed by revisiting our discussion of the CLFA model in chapter 3. By taking conditional expectation with regard to the factors from say, (3.9), for a given set of factor values and a particular observed variable, Y_s, which is assumed continuous as in the (standard) regression analysis setting, it follows that

$$\mathcal{E}(Y_s|f_1, \ldots, f_m) = \mu_s + a_{s1}f_1 + a_{s2}f_2 + \cdots + a_{sm}f_m \tag{4.3}$$

holds $(1 \le s \le p)$. Equation (4.3) will soon be seen as particularly helpful conceptually also in IRT contexts.

4.1.2 Extending the linear modeling idea to discrete response variables

As indicated above, a possible representation of the multiple regression analysis model is the following one (for a given set of predictor values, as assumed in the rest of this chapter):

$$\mathcal{E}(Y) = \beta_0 + \beta_1 X_1 + \beta_2 X_2 + \cdots + \beta_q X_q = \mu, \quad \text{say} \tag{4.4}$$

The generalized linear model (GLIM) extends this linear modeling idea from standard regression analysis to the case of response variables that are not continuous, while belonging to the so-called exponential family that includes distributions of relevance in this book (for example, Agresti [2013]). To this end, in the present setting, the GLIM postulates not the mean μ but a function of it, called a link function, to be still linearly related to a considered set of predictors as in the middle part of (4.4) (compare Raykov and Marcoulides [2011, 2013]). More concretely, denoting this function by $g(\mu)$, say, which is suitably chosen, a GLIM stipulates the following relationship (using for convenience the same notation as above):

$$g(\mu) = \beta_0 + \beta_1 X_1 + \beta_2 X_2 + \cdots + \beta_q X_q$$

That is, a GLIM posits that

$$g\{\mathcal{E}(Y)\} = \beta_0 + \beta_1 X_1 + \beta_2 X_2 + \cdots + \beta_q X_q \tag{4.5}$$

There are various choices of the link function $g(\cdot)$, discussed in detail in more advanced treatments of GLIM (for example, Dobson and Barnett [2008]). This multitude of possible link functions is what makes the GLIM framework highly comprehensive. [We mention in passing that as a special case, which results when $g(\cdot)$ is the identity function, that is, $g(\mu) = \mu$, for a normally distributed response variable, the GLIM includes also the standard regression model.] We will soon be concerned with another special case of a GLIM, which is well suited for binary or binary scored items within the IRT framework.

4.1.3   The components of a generalized linear model

A GLIM consists of the following three essential elements, or parts (for example, Agresti [2013]):

i) a random component (also referred to as the "sampling model" or "variable distribution");

ii) a link function [namely, $g(\cdot)$]; and

iii) a systematic component—the expression in the right-hand side of (4.5), also called the "linear predictor".

Element i) refers effectively to the distribution of the dependent variable, Y (compare Raykov and Marcoulides [2011]). For example, as is often the case in an IRT context when considering a single item, this variable may be dichotomous, as for a binary or binary scored item. In this case, Y follows a Bernoulli distribution, with a probability of taking the value of 1, say, being p and the value 0 with probability $1 - p$ (see chapter 3; see also De Boeck and Wilson [2015]). Element ii) refers to the link function $g(\cdot)$ that is applied on the mean of the response, Y, before its linear relationship with a set of predictors is considered or postulated within the model [see (4.4)]. When this outcome measure is binary, the optimal choice of the link function is as the logit function, that is, the inverse of the logistic function (see Dobson and Barnett [2008] and chapter 2). The logistic function option will be discussed in more detail later in this and in the next chapter within the IRT framework. Element iii) of a GLIM describes what is frequently referred to as "the linear predictor". This is a linear combination of the predictors under consideration, including an intercept [see (4.4) and surrounding discussion]. Thereby, just as in standard regression analysis, the distribution of the predictors is not of interest, that is, is left unrestricted (unstructured), because it is not modeled. In particular, in a GLIM, the explanatory variables can be measured on a nominal, ordinal, interval, or ratio scale and can be discrete or continuous. The explanatory variables participating in the "linear predictor", however, are assumed in the conventional GLIM framework, as in regression analysis, to be measured without error (for example, Agresti [2013]). We will be saying more on this assumption later in this and in the next chapter.

Generalized linear modeling was first developed in the early 1970s by Nelder and Wedderburn (1972). The research by these and other statisticians in the subsequent years has demonstrated that one can use a single numerical approach, the so-called iteratively reweighted least squares, to accomplish maximum likelihood estimation as well as related standard error evaluation and hypothesis testing. Later work by numerous other scholars has contributed substantially to the many uses that the GLIM framework has found meanwhile in applied statistics and the empirical sciences.

4.2   Logistic regression as a generalized linear model of relevance for item response theory and item response modeling

When a dependent variable is the response on a binary or binary scored item, as is oftentimes the case in applications of IRT in the behavioral, educational, and social sciences, a special case of the GLIM is appropriate to use as a means of modeling and analyzing the relationship of that variable to a given set of predictors, X_1 through X_q ($q \geq 1$; see also De Boeck and Wilson [2015]). This special case for a discrete outcome, denoted Y in this book, is typically referred to as logistic regression (LR). In the particular setting of a binary response, this regression model is frequently called the binary logistic regression (BLR) model. As we indicated earlier in the chapter, LR and BLR assume conventionally that each of the explanatory variables is measured without any error, or in practical terms, only with minimal, negligible error (for example, see Agresti [2013]; see also below).

4.2.1   Univariate binary logistic regression

As can be seen from the preceding discussion, BLR is a special case of GLIM with i) the Bernoulli distribution being the "random component", ii) the logit function being the link, and iii) the linear combination of the explanatory variables (with an intercept), namely, $\beta_0 + \beta_1 X_1 + \beta_2 X_2 + \cdots + \beta_q X_q$, being the linear predictor. Hence, the model underlying BLR is as follows:

$$\text{logit}(\mu) = \text{logit}\{\mathcal{E}(Y)\} = \beta_0 + \beta_1 X_1 + \beta_2 X_2 + \cdots + \beta_q X_q \tag{4.6}$$

That is, the BLR model is [see also (4.4) and chapter 2]

$$\ln\frac{\mu}{1-\mu} = \beta_0 + \beta_1 X_1 + \beta_2 X_2 + \cdots + \beta_q X_q$$

Alternatively, because $\mu = \mathcal{E}(Y) = p$ is the probability of response 1 on the binary outcome Y, as pointed out in chapter 3, from (4.6), it follows that this model can be reexpressed as

$$\ln\frac{p}{1-p} = \beta_0 + \beta_1 X_1 + \beta_2 X_2 + \cdots + \beta_q X_q \tag{4.7}$$

($p < 1$; in the last three equations, strictly speaking, one considers the predictors as given, as indicated earlier). We stress that no error term appears at the end of the right-hand side of (4.7). The reason is that on the left side of this equation, we have a function only of the mean of the outcome. Hence, an error term has been effectively already accounted for (with a mean of 0) in that equation. Further, we note that in the remaining chapters, for convenience, we will refer to the $\beta_0, \beta_1, \ldots, \beta_q$ parameters in this model as intercept and slopes (partial LR slopes), respectively. We prefer to do this also to retain a reference similar to that used in standard regression analysis (for example, Raykov and Marcoulides [2013]).

From (4.7) follows yet another, very useful, and equivalent representation of the binary logistic model in terms of what may be called "model implied probability" for the event under consideration, for example, "correct" response (denoted or scored "1"). This model representation is obtained with straightforward algebraic rearrangements from (4.7),

$$
\begin{aligned}
p &= \frac{\exp(\beta_0 + \beta_1 X_1 + \cdots + \beta_q X_q)}{1 + \exp(\beta_0 + \beta_1 X_1 + \cdots + \beta_q X_q)} \\
&= \Lambda(\beta_0 + \beta_1 X_1 + \cdots + \beta_q X_q)
\end{aligned}
\tag{4.8}
$$

where $\Lambda(\cdot)$ is the earlier discussed logistic function (see also chapter 2; as mentioned before, $\exp(\cdot)$ is used to denote exponentiation throughout the book). The following reexpression of (4.8) will be of special importance to us in the IRT- and IRM-related discussions in the remainder of the book [note that the numerator in the middle part of (4.8) is always positive and hence can be canceled],

$$
\begin{aligned}
p &= \frac{1}{1 + \exp\{-(\beta_0 + \beta_1 X_1 + \cdots + \beta_q X_q)\}} \\
&= [1 + \exp\{-(\beta_0 + \beta_1 X_1 + \cdots + \beta_q X_q)\}]^{-1} \\
&= \frac{1}{1 + \exp\{-\mathrm{logit}(p)\}}
\end{aligned}
\tag{4.9}
$$

where $\mathrm{logit}(p) = \beta_0 + \beta_1 X_1 + \cdots + \beta_q X_q$ denotes the logit of the probability p [see chapter 2 and the above (4.6) and (4.7)].

A particularly interesting special case of this discussion that will be useful for our aims in the rest of the book is rendered when $k = 1$. In this case, only one independent variable, say, X_1, is used for the purpose of explaining individual differences in the binary response variable Y. Then the BLR model, in the form of (4.9), is as follows:

$$
\begin{aligned}
p &= \frac{\exp(\beta_0 + \beta_1 X_1)}{1 + \exp(\beta_0 + \beta_1 X_1)} \\
&= \Lambda(\beta_0 + \beta_1 X_1) \\
&= \frac{1}{1 + \exp\{-(\beta_0 + \beta_1 X_1)\}}
\end{aligned}
\tag{4.10}
$$

In other words, the model implied probability is then the logistic function. In particular, its argument is a linear function of the predictor, using the intercept and LR slope, namely, $\beta_0 + \beta_1 X_1$. [This argument is the expression within parentheses in the denominator of the right-most side of (4.10), that is, equals the logit of p; see chapter 2.] Hence, upon estimation of the parameters, the model predicted probability is

$$
\widehat{p} = \frac{1}{1 + \exp\left\{-\left(\widehat{\beta}_0 + \widehat{\beta}_1 X_1\right)\right\}}
$$

where a caret is used to denote estimate (estimator) of the quantity underneath. In other words, if a particular subject's value on the predictor is X_{1i}, then his or her predicted probability, $\widehat{p}_i$, of "correct" response according to this model is

$$\widehat{p}_i = \frac{1}{1 + \exp\left\{-\left(\widehat{\beta}_0 + \widehat{\beta}_1 X_{1i}\right)\right\}} \quad (1 \leq i \leq n)$$

The typical approach to estimating the BLR model (and the LR model more generally) is via use of maximum likelihood (compare Cai and Thissen [2015]). Thereby, the resulting maximum-likelihood estimators possess highly desirable statistical properties with large samples, such as unbiasedness, consistency, normality, and efficiency (for example, Agresti [2013]). In addition, these estimators have the property of invariance and represent in general functions of sufficient statistics (for example, Casella and Berger [2002]). For more details related to LR, as well as a discussion on testing and comparison of LR models, reference is made for instance to Hosmer, Lemeshow, and Sturdivant (2013).

4.2.2 Multivariate logistic regression

A multivariate version of BLR (and more generally of LR with a discrete response) is obtained directly from (4.7). To this end, we include in it, say, k binary outcome variables ($k > 1$), denoted Y_1 through Y_k, and study simultaneously their relationship to the q predictors used, X_1 through X_q. Also, here a standard assumption is that each of the latter variables is measured free of error, or, in practical terms, with minimal error that is negligible (compare Dobson and Barnett [2008]). Specifically, the multivariate BLR model is then as follows:

$$\ln\frac{\mu_1}{1 - \mu_1} = \beta_{01} + \beta_{11}X_1 + \cdots + \beta_{q1}X_q$$

$$\ln\frac{\mu_2}{1 - \mu_2} = \beta_{02} + \beta_{12}X_1 + \cdots + \beta_{q2}X_q$$

$$\vdots$$

$$\ln\frac{\mu_k}{1 - \mu_k} = \beta_{0k} + \beta_{1k}X_1 + \cdots + \beta_{qk}X_q \tag{4.11}$$

In (4.11), $\mu_j = \mathcal{E}(Y_j) = p_j$ is the probability of "correct" response on Y_j (by someone with given values on the explanatory variables), and β_{0j} and β_{sj} are, respectively, the intercept and partial LR coefficients associated with the jth outcome variable. A more compact representation of (4.11) is as

$$\text{logit}(p_j) = \beta_{0j} + \beta_{1j}X_1 + \cdots + \beta_{qj}X_q$$

$[j = 1, \ldots, k, s = 1, \ldots, q;$ see also (4.6) and (4.7)].

The version of the multivariate BLR model (4.11), which results when $k = 1$, will be of particular conceptual interest to us in the remainder of this book. In that case, the multivariate binary logistic model is

$$\ln\frac{\mu_1}{1 - \mu_1} = \beta_{01} + \beta_1 X$$

$$\ln\frac{\mu_2}{1 - \mu_2} = \beta_{02} + \beta_2 X$$

$$\vdots$$

$$\ln\frac{\mu_k}{1 - \mu_k} = \beta_{0k} + \beta_k X \qquad (4.12)$$

That is, according to this BLR model, the k log odds for "correct" response on the binary or binary scored items Y_1 through Y_k are functions of the observed predictor X. We stress that this model includes two sets of k parameters each, which consist of the intercepts and slopes, respectively, and play the role here correspondingly of location and scale parameters for the individual items (compare Hosmer, Lemeshow, and Sturdivant [2013]; see also chapter 2). Equivalently, the logits of the probabilities of "correct" response on Y_1 through Y_k are functions of these two sets of parameters, as stated in (4.12). (Their counterpart parameters in the case of an unobserved predictor will soon become of special interest to us; see chapter 5.)

As noted earlier in this section, with straightforward algebraic rearrangements, we can obtain from (4.12) the following important equivalent form of the multivariate BLR model. This form will be of particular conceptual aid to us in subsequent chapters of the book,

$$p_1 = \frac{1}{1 + \exp\{-(\beta_{01} + \beta_1 X)\}}$$

$$p_2 = \frac{1}{1 + \exp\{-(\beta_{02} + \beta_2 X)\}}$$

$$\vdots$$

$$p_k = \frac{1}{1 + \exp\{-(\beta_{0k} + \beta_k X)\}}$$

where $p_1, \ldots, p_k$ are the probabilities of "correct" response on the k binary (or binary scored) items of concern here.

The reason this section was concerned with LR, and especially with (multivariate) BLR, is that several popular IRT models can be considered analogs of multivariate BLR models for a given set of k binary or binary scored items (the above Y_1 through Y_k). While this analogy will be easy to see, it will be important to remember that it cannot be carried through completely in an empirical setting. This is because the predictor (or predictors) in a corresponding IRT model is not directly observed like any of the above explanatory variables X_1 through X_q are in conventional or standard LR models. As we will elaborate further in the next chapter, we are dealing in unidimensional IRT with a single predictor as well (see also chapter 1). However, unlike any of the predictors $X_1, \ldots, X_q$ here, that one will not be directly observable and only possible to measure indirectly and imprecisely in the observed or recorded binary item responses. Before we discuss the pertinent details though, it will be beneficial to highlight another related connection of the GLIM to a main part of latent variable modeling that we will see soon as closely linked to most popular IRT models.

4.3 Nonlinear factor analysis models and their relation to generalized linear models

4.3.1 Classical factor analysis and its connection to generalized linear modeling

We begin by recalling the earlier discussed CLFA model (see chapter 3 and notation used there):

$$\mathbf{Y} = \boldsymbol{\mu} + \mathbf{Af} + \mathbf{u} \tag{4.13}$$

In (4.13), $\mathbf{Y}$ symbolizes a given vector (set) of observed continuous variables for which one seeks in CLFA one or more underlying traits, far fewer in number usually and collected in the vector $\mathbf{f}$, which explain to a satisfactory degree the interrelationships among the measures in $\mathbf{Y}$. We mention in passing that for several decades in the early history of FA, the assumption $\boldsymbol{\mu} = \mathbf{0}$ was nearly automatically made by its users. This assumption effectively ignores the origin of the scale on which the observed variables $Y_1, \ldots, Y_k$ are measured and was made for convenience reasons because it does not affect the typically analyzed then observed correlation matrix. In the general case, however, $\boldsymbol{\mu} = \mathbf{0}$ is not an assumption of the FA model. In fact, the real benefit of considering the general FA model with a nonzero intercept vector $\boldsymbol{\mu}$ will soon be seen when realizing that an intercept-like parameter is an essential part also of popular IRT models.

This relationship between the general FA framework and IRT is best revealed if one adopts the framework of the GLIM, as we do in the remainder of this section (compare Raykov and Marcoulides [2011]). To this end, first we note from (4.13) that the expectation of any observed variable Y_j, given values of the m factors, results as

$$\mathcal{E}(Y_j) = \mu_j + a_{j1}f_1 + a_{j2}f_2 + \cdots + a_{jm}f_m \tag{4.14}$$

$(j = 1, \ldots, k)$. That is, the expected score on the manifest measure Y_j is a linear function then of the factors under consideration (including an intercept, as is the case typically in regression analysis or the general linear model). We note next that the k equations in (4.14) can be written in complete form as

$$\mathcal{E}(Y_1) = \mu_1 + a_{11}f_1 + a_{12}f_2 + \cdots + a_{1m}f_m$$
$$\mathcal{E}(Y_2) = \mu_2 + a_{21}f_1 + a_{22}f_2 + \cdots + a_{2m}f_m$$
$$\vdots$$
$$\mathcal{E}(Y_k) = \mu_k + a_{k1}f_1 + a_{k2}f_2 + \cdots + a_{km}f_m$$

Compactly, the last set of k equations is reexpressed as

$$\mathcal{E}(\mathbf{Y}) = \boldsymbol{\mu} + \mathbf{A}\mathbf{f} \qquad (4.15)$$

with the same notation as in chapter 3. Equations (4.15) in effect say that the expectations of the observed variables are linearly related to the underlying factors (at the given values for them).

As indicated earlier, a special case of particular interest in many measurement contexts in the behavioral, educational, and social sciences is obtained from (4.15) when $m = 1$ factor is used in them. In that case of unidimensionality, at times referred to also as homogeneity, (4.15) reduces to

$$\mathcal{E}(\mathbf{Y}) = \boldsymbol{\mu} + \mathbf{a}f \qquad (4.16)$$

where f denotes the single factor of relevance and $\mathbf{a}$ symbolizes the column vector of the observed variable loadings (weights) on it.

With the preceding discussion in this subsection, we are now ready to extend our vision of FA to nonlinear FA as generalization of CLFA.

4.3.2   Nonlinear factor analysis models

The conceptual idea underlying the general CLFA model in (4.14) with continuous manifest measures is helpful also with other types of item responses. These equations, and in particular their consequence (4.16) for the case of a single factor, relate in general a dependent variable to one or more explanatory variables. That relationship idea is in fact equally beneficial when the manifest measures Y_j are binary variables as in this chapter ($j = 1, \ldots, k$; compare Bartholomew [1996]). To this end, we can make use of the general modeling idea underlying the comprehensive GLIM framework. Specifically, recalling that the expected observed score is the probability of "correct" response (denoted "1"), we see that the GLIM approach readily provides the insight needed. This is achieved, for instance, via the logit link function, as mentioned before. We assume thereby that one or more latent factors underlie individual differences on an examined item, say, Y_j, and hence can postulate that

$$\text{logit}\{\mathcal{E}(Y_j)\} = \mu_j + a_{j1}f_1 + a_{j2}f_2 + \cdots + a_{jm}f_m \qquad (4.17)$$

[$j = 1, \ldots, k$; see (4.6); we use in the right-hand side of (4.17) formally the same notation as in (4.14) to emphasize their conceptual relation, rather than to imply identity of the corresponding individual terms in these two sets of equations].

We easily observe that the right-hand sides of (4.17) are linear functions of unobserved variables, the postulated factors (given values of them). Specifically, the logits in (4.17) are stipulated as linear in the parameters involved, the μ's and the a's. In fact, (4.17) can be viewed as representing an FA model for the logits of the probabilities of correct response for each of the observed measures. We underscore though that (4.17) is merely an extension of the FA model (4.15), which we have obtained using the general modeling idea underlying the GLIM framework. To emphasize this conceptual link between FA and GLIM, simultaneously all p (4.17) can be written as follows (for example, Raykov and Marcoulides [2011]):

$$\text{logit}\{\mathcal{E}(Y_1)\} = \mu_1 + a_{11}f_1 + a_{12}f_2 + \cdots + a_{1m}f_m$$
$$\text{logit}\{\mathcal{E}(Y_2)\} = \mu_2 + a_{21}f_1 + a_{22}f_2 + \cdots + a_{2m}f_m$$

$$\vdots$$

$$\text{logit}\{\mathcal{E}(Y_k)\} = \mu_k + a_{k1}f_1 + a_{k2}f_2 + \cdots + a_{km}f_m \qquad (4.18)$$

or, in matrix form,

$$\mathbf{logit}\{\mathcal{E}(\mathbf{Y})\} = \boldsymbol{\mu} + \mathbf{A}\mathbf{f} \qquad (4.19)$$

[with corresponding notation, formally the same as in (4.15); see above]. In (4.19), $\mathbf{A} = [a_{ij}]$ is the matrix of weights or loadings, $\boldsymbol{\mu} = (\mu_1, \mu_2, \ldots, \mu_k)'$ is the vector of pertinent intercepts, and the symbol $\mathbf{logit}\{\mathcal{E}(\mathbf{Y})\} = \mathbf{logit}(\mathbf{p})$ stands for the $k \times 1$ vector of logits of the expectations or means, or probabilities of "correct" response, of the items Y_1 through Y_k. That is, (4.19) can be said to define an FA model for the logits of the probabilities of correct response on the studied set of binary or binary scored items, as indicated above.

We observe easily that the right-hand side of (4.18) is "identical" to that side of the CLFA model in (4.14) (see above remark on used notation). Further, the left-hand side of (4.18) is a nonlinear function of the observed variable mean (expectation) vector, rather than just stating that mean (compare CLFA model). With this in mind, (4.18) [and (4.19)] can actually be viewed as defining a nonlinear FA model that is based on the logit link (for example, Raykov and Marcoulides [2011, 2016b]; see also McDonald [1999]).

Alternatively, the used general framework provides just as well the opportunity to connect the "linear predictor", that is, the right-hand side of (4.14), with a discrete response variable via the probit link function. (As we saw in chapter 2, this function is essentially identical to the logistic function after a rescaling that has no practical consequences most of the time in IRT contexts; compare Raykov and Marcoulides [2011].) Specifically, using for convenience the same notation as in (4.18), we see that

$$\text{probit}\{\mathcal{E}(Y_j)\} = \Phi^{-1}\{\mathcal{E}(Y_j)\} = \mu_j + a_{j1}f_1 + \cdots + a_{jm}f_m \qquad (4.20)$$

holds ($j = 1, \ldots, k$). By complete analogy to the preceding discussion in this section leading up to the nonlinear FA model based on the logit link, compactly (4.20) is written as

$$\mathbf{probit}\{\mathcal{E}(\mathbf{Y})\} = \boldsymbol{\mu} + \mathbf{A}\mathbf{f} \qquad (4.21)$$

where $\mathbf{probit}\{\mathcal{E}(\mathbf{Y})\}$ stands for the $k \times 1$ vector of probits of the expectations or means (probabilities of "correct" response) of the individual items in $\mathbf{Y}$. Hence, (4.21) can be viewed as defining a nonlinear FA model based on the probit link. In particular, (4.21) can be seen as an FA model for the probits of the probabilities of correct response for the studied set of binary or binary scored items.

Equations (4.19) and (4.21) share one important feature. This is the fact that they are based on a Bernoulli sampling model. This model reflects the random mechanism behind a binary random variable, the response on a corresponding item of interest (compare Raykov and Marcoulides [2011]). In addition, these two sets of equations are based each on a particular link function. The latter is applied on the probability of observing correct response on the items $Y_1, \ldots, Y_k$, which is the mean of each of them (see chapter 3). Furthermore, in the right-hand sides of these two sets of equations, the same linear predictor is present. Indeed, this is a linear function of the same set of assumed underlying traits (factors), denoted $f_1, f_2, ..., f_m$. Hence, the two nonlinear FA models defined by (4.19) and (4.21) could also be viewed more broadly as GLIMs (see the following for a qualification).

Having observed that, we now pay attention to the fact that participating in the right-hand sides of these two GLIM-based extensions of the general FA model are actually unobserved variables, the m latent traits. Thus, one can refer to the two nonlinear FA models in (4.19) and (4.21) as generalized latent linear models (for example, Bartholomew, Knott, and Moustaki [2011]). The extension of generalized latent linear models to the mixed modeling context, for example, the multilevel setting (where studied persons are nested in higher-order units, such as schools, hospitals, physicians, teams, companies, or cities), is referred to as the generalized latent linear and mixed model (see, for example, Skrondal and Rabe-Hesketh [2004]; see also De Boeck and Wilson [2015]). Their very wide and comprehensive framework goes beyond the confines of this book and can be pursued in more advanced treatments of latent variable modeling (for example, Skrondal and Rabe-Hesketh [2004]).

4.4   Chapter conclusion

In this chapter, we have accomplished our preparation for discussing popular IRT models, which we will deal with in the next chapter. Summarizing the developments in the present one, we see that the used comprehensive GLIM framework has proved highly useful in i) providing us with the multivariate (binary) LR framework as a special case of GLIM, which will be a valuable aid to us when discussing IRT models in the following chapters; and ii) allowing us also to extend the CLFA model to more general settings that are of relevance when an empirical researcher is facing discrete items with few response options, as is often the case in the behavioral, educational, and social disciplines. The GLIM framework of this chapter has also permitted us to see how one can extend the CLFA model to nonlinear FA models, which will be helpful in the subsequent chapters of the book. We will use these important statistical framework connections there in the specific context of widely used IRT models.

5 Fundamentals of item response theory and item response modeling

5.1 Item characteristic curves revisited

As was first mentioned in chapter 1, a central concept in (unidimensional) item response theory (IRT) is that of the item characteristic curve (ICC). Its definition is based on the assumption that there is an underlying latent trait (ability, construct, or latent variable) that we are willing to study but cannot directly measure. This trait or ability underlies the examined persons' performance on a given set of items or measuring instrument of interest. However, their exact locations on that unobserved continuum are unknown. Using the available data on the items (presumably) tapping into that latent dimension, we wish to make inferences about i) the characteristics of the items and ii) the unknown positions of the persons on the unobserved continuum.

As we indicated earlier in the book, until chapter 11, we assume that we are dealing with binary or binary scored items. We also assume until chapter 12 a single (unidimensional) underlying trait or ability that is evaluated by them. In this widely used setting in the behavioral, educational, and social sciences, the discussion in chapter 1 on the ICC allows us to interpret further this fundamental IRT concept. As will be recalled, for a given item, its ICC was defined as the probability $P(\theta)$ of "correct" response on it by a person with a level (value) of θ on the studied trait or ability. [An alternative and equivalent formal representation of this probability is as $P(Y = 1|\theta)$, where Y is as before the random variable (RV) representing the item response and $P(\cdot|\cdot)$ denotes conditional probability, given θ (for example, van der Linden [2016b]); for simplicity, we will use the former notation $P(\theta)$ in the remainder of the book.] Therefore, as such a probability, the ICC is actually the regression, or conditional expectation, of the observed dichotomous item score Y upon the underlying latent dimension θ (Casella and Berger 2002). That is,

$$\text{ICC} = P(\theta) = \mathcal{E}(Y|\theta) \tag{5.1}$$

holds, where $\mathcal{E}(.|.)$ denotes conditional expectation of the observed item score (given θ). Equation (5.1) explicates formally the fact that the ICC is a function of θ and represents the former as a (conditional) mean of the response variable (for example, Reckase [2009]).

5.1.1 What changes across item characteristic curves in a behavioral measurement situation?

The preceding discussion is not meant to imply that in IRT and item response modeling (IRM) all items in a given set or measuring instrument are expected to have the same ICC. The discussion similarly does not mean to suggest that any application of IRT needs to assume the same functional relationship between the probability of "correct" response and the studied trait, ability, or latent variable (see also chapter 11).

To further clarify this matter, as alluded to earlier in the book, different IRT models postulate different functions for $P(\theta)$ and thus are associated with different ICCs. In addition, even for a given model, the items are usually assumed to have ICCs that differ in their characteristics. More specifically, for a chosen IRT model, all items are most of the time presumed to have ICCs that come from the same functional class while having one or more parameters with varying values across them. That is, once an IRT model is decided for (see, for example, the next chapter), usually all items in a considered set are stipulated to have ICCs that are described with the same functional form, that is, represent the same function of θ. However, their ICCs have different particular features, that is, differ in what we will be referring to as item parameters, whose values are item specific.

To be more concrete, we see that typical IRT models differ from one another in i) the assumed class of functions describing the ICCs (usually, but not always, with one function per IRT model) and ii) the number of parameters on which these functions depend. Further, once an IRT model is chosen, items usually differ from each other in the values that the parameters mentioned in ii) take. In particular, each item is characterized by a specific set of numerical values for these parameters. This assumption of identical functional form is relaxed in more general IRT models, which are called hybrid models. In these models, different items may have ICCs from distinct functional classes. Hybrid models are currently less frequently used in empirical studies using IRT but are just as applicable, and will be discussed later in the book (see chapter 11).

However, the ICCs in an IRT model of interest in this book (for binary and binary scored items) have a very important property in common. This is the feature that as one approaches the "central" part of the ICC, small changes in ability lead to more pronounced changes in the probability of correct response than is the case further away from that central part (see, for instance, figure 1.1 in chapter 1). Specifically, in the "middle" of the central part of the ICC, the gradient (steepness or tilt) of this curve is highest or most pronounced. (See also later sections of this chapter and its footnote 1.) Moreover, as one moves from one end of the central part of the ICC to the other end of it, the ICC is actually positioned at different sides of the pertinent tangent or gradient line at the point that is right "in the middle" of the ICC. This is because at that location, in the middle of its central part, as indicated earlier in the book, an ICC has an inflection point (its second derivative, as a function of θ, vanishes; see also figure 2.2 in chapter 2 as well as its section 2.1 for graphical illustration and discussion of the inflection point notion).

In addition to this property of differentiation between subjects with distinct values on the underlying trait or ability, as indicated previously in the book, all items in a typical IRT model share another important feature. This is the property of local independence (LI), which has an important relationship to that of unidimensionality (or homogeneity) of a measuring instrument or item set under consideration. Thus, we discuss next in more detail their relationship.

5.2 Unidimensionality and local independence

In many empirical applications of statistics and behavioral measurement, it is common to talk about relatedness or unrelatedness, for example, correlation or lack thereof between RVs of interest (compare Raykov and Marcoulides [2011]). In IRT and its applications, it turns out that it is more beneficial to discuss matters in terms of statistical dependence or, alternatively, in terms of statistical independence. Before we proceed further, let us recall that uncorrelatedness and independence are in general two different notions when considering, say, two or more continuous RVs. In particular, lack of correlation coincides with independence only when an additional condition holds, namely, normality of the RVs (for example, Casella and Berger [2002]). Further, for two RVs, independence implies lack of correlation, but the reverse is only true when the variables are bivariate normal, and hence univariate normal as well (for example, Raykov and Marcoulides [2008]).

With discrete RVs, such as for instance the responses on two or more dichotomous items in a measuring instrument, the notion of item independence can be obtained from a more general definition of independence of RVs. Accordingly, two RVs are independent if their distribution when considered simultaneously, called joint distribution, can be represented in terms of their individual distributions, called marginal distributions. More concretely, the RVs are independent if their event probability when taken together is the product of the individual variable probabilities for the pertinent individual events (see next and, for example, Casella and Berger [2002]).

When one uses discrete RVs, and in particular binary responses of relevance for the majority of the chapters in this book, the independence definition can be specialized further as follows (for example, Raykov and Marcoulides [2011]). To this end, let us first denote by $P_j(+)$ the probability of "correct" response on item j; by $P_j(-)$ that of "incorrect" response on it; by $P_{jr}(+,+)$ the probability of obtaining "correct" responses both on the jth and rth items ($1 \leq j, r \leq k; j \neq k$); and correspondingly define the notation for the probability of other types of responses for a given pair in a set of k items of interest ($k > 1$). Then the jth and rth items are called (statistically) independent if and only if

$$P_{jk}(+,+) = P_k(+)P_j(+)$$
$$P_{jk}(+,-) = P_k(+)P_j(-)$$
$$P_{jk}(-,+) = P_k(-)P_j(+)$$
$$P_{jk}(-,-) = P_k(-)P_j(-) \tag{5.2}$$

hold. Conversely, if one or more of the four equations in the set (5.2) is not fulfilled, the two items are called (statistically) dependent. In simple terms, two binary or binary scored items are independent if the probability of any pair of responses on them is obtainable by multiplying the probabilities for the pertinent responses on the individual items.

In general, we will call two RVs statistically dependent when they are not statistically independent. For a given set of dichotomous items, for instance, statistical independence implies that the response on one of them has nothing to do with that on any other item. The items will be dependent, however, if the responses on them are related. In particular, when constructing a measuring instrument presumably evaluating a single underlying trait, construct, or ability, as could be implied from figure 1.2 in chapter 1, one would want to use items that are dependent. Specifically, their dependence would be expected to result from the fact that they evaluate a common latent dimension or continuum. When this is the case indeed, that is, a considered set of items evaluate just one underlying latent dimension, they are referred to as unidimensional or alternatively as homogeneous items (item set or measuring instrument).

5.2.1   What are the implications of unidimensionality?

The concept of unidimensionality, as alluded to in earlier chapters, is of special interest for the aims of this book and more generally in behavioral, educational, and social measurement (compare Raykov and Marcoulides [2011]). If this strong assumption is fulfilled, it yields many important benefits, particularly when interpreting observed scores resulting from a homogeneous item set or instrument. It is these types of benefits that one can also capitalize on when conducting IRM in an empirical setting or more generally carrying out measurement in these and related disciplines.

As was indicated above, a set of items is unidimensional if there is only one trait, construct, or ability that can be used to explain the interrelationships among them. That is, in such cases, there is a single trait or ability that accounts for the statistical dependence among the items. In conceptual terms, therefore, unidimensionality in IRT means essentially the same as unidimensionality or the assumption of a single underlying factor in a factor analysis model. We stress that unidimensionality is defined as an overall population notion. Indeed, it refers to the (presumed) existence of a single trait or ability that explains the lack of item independence in a studied subject population. That is, unidimensionality is consistent with the items being statistically dependent in a population of concern.

In this context, a natural question that arises is the following: What happens with a unidimensional instrument (item set) if we "fix" the underlying trait, construct, or ability, that is, θ? In other words, if we consider only a "slice" of the population in question, which has the property that every person in the slice possesses the same value of θ, what would be the relationship between the items in the instrument for these subjects?

Because unidimensionality means, as indicated above, that there is only one trait at work in the entire population (as far as a given set of items is concerned), it follows that in the focused "slice" of the population, there are no further abilities, traits, or constructs that could produce any joint variability in the items. Hence, the latter are no longer dependent, that is, are statistically independent (in that subpopulation). In other words, conditional on the value of θ—which means we consider only the population "slice" mentioned—under unidimensionality, the considered items are independent. Thus, the items are locally independent. That is, LI follows from unidimensionality.

An interesting query in this connection is whether the reverse of the last statement is also correct? The answer, however, is no. Indeed, as pointed out when we discussed particular aspects of LI earlier in the book in the context of factor analysis (see chapter 4), LI does not require unidimensionality per se. In fact, LI merely means observed variable (item) independence after one fixes all underlying traits or abilities—regardless of their number—that "produce" or are responsible for the variable interrelationships in the population at large. In other words, unidimensionality is sufficient for LI, but LI is necessary for unidimensionality (compare Raykov [2011]; see below).

At this point, it is worthwhile digressing briefly to highlight the distinction between the widely used terms "sufficient condition" and "necessary condition" in mathematics and logic, which are also of special relevance for other sciences and applied disciplines. The reason is that it will be very useful in the remainder of the book, as well as in empirical applications of IRT, to be in a position to distinguish between different conditions (statements) and their logical relationships. Specifically, we define a condition A as sufficient for another condition B if the validity of A brings about or implies the validity of B. Alternatively, a condition C is called necessary for a condition D if D cannot be valid unless C is so. (Note that when C is necessary for D, it may be that additional conditions beyond C also need to be fulfilled for D to be valid; see, for example, Raykov [2011] and the next example.) For instance, let us ask the question when an integer number is divisible by 4. As can be readily found out given the preceding definition, for an integer number, divisibility by 4 is sufficient for it to be even (divisible by 2). The reason is that divisibility by 4 implies that a number is even because there is no odd number that is divisible by 4. However, being even is only necessary for a number to be divisible by 4. This is because not every even number is divisible by 4. In other words, just being even is not enough or does not guarantee or suffice for a number to be divisible by 4. (In fact, a sufficient condition for a given number a to be divisible by 4 is that the number b made of its last 2 digits is itself divisible by 4, that is, b being divisible by 4. [If a has just 1 digit, place a zero before it to obtain b for the purpose of this argument; if a has 2 digits, take $b = a$.]) That is, one needs to request "more" from *a* than just being even for *a* to be divisible by 4. As this simple example illustrates, oftentimes for a given condition of interest, say, *K*, it is more difficult to find a condition that is sufficient for *K* than it is to come up with a necessary condition for *K*.

Graphically, we can represent this consequential logical relationship between the notions of unidimensionality and LI using figure 5.1, reminiscent of a Venn diagram (for example, Roussas [1997]). The figure displays two sets corresponding to unidimensionality and LI, and it is important to note that the set for the former concept is positioned within that for the latter notion. (Other than that, no particular meaning is to be attached to the shape of the outside and inside contours.) With the preceding discussion in mind, one sees from figure 5.1 that if LI does not hold, then neither does unidimensionality. Moreover, one also sees that unidimensionality implies LI, as we have already seen earlier in this section. However, LI alone does not imply unidimensionality. That is, when LI holds, then unidimensionality may or may not hold. Hence, based also on figure 5.1, one sees that i) LI is a necessary condition for unidimensionality and that ii) unidimensionality is a sufficient condition for LI. (The figure is meant to imply that under unidimensionality, one needs to fix the underlying trait or ability to obtain LI.)

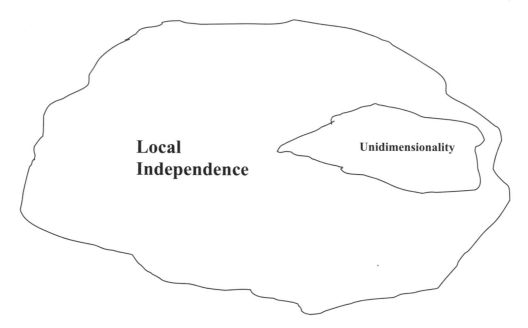

Figure 5.1. Relationship between the concepts of local independence and unidimensionality

This discussion brings us to the following definition of the critically important concept of dimensionality for this book and IRT discussions and applications (compare Raykov and Marcoulides [2011]). Accordingly, dimensionality of an item set or a measuring instrument of interest is defined as the minimal number of latent traits needed to achieve LI, that is, statistical independence of the items, after fixing or conditioning on these traits. Hence, if for a given set of items, m is the smallest integer number $(m > 0)$ with the property that m latent traits are needed to be taken into account (fixed or conditioned on) to achieve LI—but any fewer than m would not accomplish it—then this item set is m-dimensional. As indicated previously in the book, of particular interest to us in all its chapters but the last is the case of $m = 1$, typically referred to as unidimensionality or alternatively as item set or measuring instrument homogeneity.

Regardless of what the value m of latent variables is in an empirical setting of application of the comprehensive latent variable modeling methodology, and in particular of IRT, we note here the following realization. One or more latent dimensions evaluated by an instrument may or may not be the same as the trait, construct, or ability of actual interest to a researcher that he or she is willing, planning, or intending to measure with a given multi-item instrument (for example, Hambleton, Swaminathan, and Rogers [1991]). As an example, just like a scale (composite) may have poor validity even though it may be homogeneous or unidimensional, so can also a set of items or an instrument in question measure a dimension (or dimensions) that is distinct from the one of substantive concern or interest. In particular, for the case of instrument unidimensionality (homogeneity), we stress that an instrument may poorly measure a trait, ability, or construct of actual interest, even though it may be assessing just one latent dimension. That is, being unidimensional for a given item set or instrument is not sufficient to claim measuring with it an ability or trait of actual interest to a researcher, because he or she may in fact be evaluating with the instrument a single dimension that is distinct from that ability or trait if not entirely different from it.

5.2.2   A formal definition of local independence

With the preceding descriptive discussion in mind, we are now ready to give a formal definition of the particularly important concept of LI for IRT and IRM. As indicated earlier (for example, chapter 2), this concept is also of special importance for other fields of latent variable modeling, such as factor analysis, latent class analysis, and latent profile analysis (finite mixture analysis).

To proceed with this definition, assume that a set of k items $Y_1, Y_2, \ldots, Y_k$ are given that are of interest to examine with respect to LI (compare Raykov and Marcoulides [2011]; $k > 1$). They are called locally independent, if for a given (fixed) set of values on m latent traits $(0 < m \le k)$, denoted $f_1, \ldots, f_m$, the joint probability of any response pattern on the k items is the product of the probabilities of pertinent response on any of the items at that set of values for the traits. More formally, this item set possesses the LI property (and the items are locally independent) if

$$
\begin{aligned}
P(Y_1 = y_1, Y_2 = y_2, \ldots, Y_k = y_k | f_1, \ldots, f_m) &= P(Y_1 = y_1 | f_1, \ldots, f_m) \\
&\times P(Y_2 = y_2 | f_1, \ldots, f_m) \ldots P(Y_k = y_k | f_1, \ldots, f_m)
\end{aligned}
\tag{5.3}
$$

where $P(.|.)$ denotes conditional probability given the condition stated after the vertical bar (for example, Raykov and Marcoulides [2013]).

We stress that in the LI definition [see (5.3)] only conditional probabilities are involved. Indeed, the LI does not include or refer to any unconditional probability. Hence, whether LI is fulfilled or not, says nothing about i) any joint probability $P(Y_1 = y_1, Y_2 = y_2, \ldots, Y_k = y_r)$ $(2 \leq r \leq k)$ and ii) any of the marginal probabilities $P(Y_1 = y_1), P(Y_2 = y_2), \ldots,$ and $P(Y_k = y_k)$. In fact, LI relates in a sense to these kinds of probabilities only after fixing the m latent traits presumably underlying subject performance on the set of items under consideration, $Y_1, Y_2, \ldots, Y_k$ (meaning that no further conditioning is necessary). We also emphasize that in the above LI definition, the traits can be continuous or categorical, as the observed variables $Y_1, Y_2, \ldots, Y_k$ can be. This realization helps us define in this way the LI concept in factor analysis, IRT, latent class, and latent profile analysis. This underscores the importance of the LI notion for a number of widely used latent variable approaches in the behavioral, educational, and social sciences and well beyond them. This fact presents LI as one of the unifying concepts for these methodologies (see also Bartholomew, Knott, and Moustaki [2011]). We conclude the present subsection by pointing out, as alluded to earlier, that LI can also be called conditional independence, as is often done in more traditional statistical literature when fixing observed variables—as "counterparts" for these purposes of the above traits $f_1, f_2, \ldots, f_m$.

5.2.3   What does it mean to assume local independence in an item response theory setting?

Given that LI is very often advanced as an assumption in an IRT setting, as well as throughout this book (unless otherwise stated or implied), it is important to keep in mind what the consequences of this assumption are (compare Reckase [2016]). Based on the preceding discussion, LI in any empirical or theoretical discussion means that the researcher assumes he or she has identified all sources of latent variability and covariability that underlie a set of observed items of interest. This implies that he or she has i) completely accounted for the dimensionality of the latent space, that is, the space spanned by all sources of latent variability (latent dimensions); and ii) included in the subsequently used IRT models all relevant unobserved dimensions that underlie subjects' performance on the item set or measuring instrument in question. This usually can be reasonably well done based on sufficiently thorough knowledge of the way the instrument functions in the population of concern and on the results of prior analyses of exploratory types (on other, independent samples) to determine the dimensionality of the trait or ability vector θ (see, for example, Raykov and Marcoulides [2011]; see also chapter 12). Hence, it would then be possible to advance the LI assumption only when one has trustworthy and complete knowledge about the dimensionality of the underlying latent space. Therefore, anytime we refer to the LI assumption in the rest of this book, we will in effect presume as usual in the literature that in the pertinent setting, one is completely aware of all latent dimensions (sources of observed variability and covariability) that underlie a studied item set or measuring instrument. A scientist must then consider subsequently only models for that set that explicitly include all these sources as separate albeit possibly interrelated dimensions. (For further discussion of this complex and consequential issue, see chapter 12.)

5.3 A general linear modeling property yielding test-free and group-free measurement in item response modeling

An important feature of the widely used general linear model, and in particular of frequently used uni- and multivariate regression models in empirical behavioral and social research, is the following somewhat underappreciated fact (compare van der Linden [2016b]). For a given set of response and explanatory variables, suppose we knew the regression model that holds exactly in a population of interest. Can we then say anything about the applicability of this model also when subpopulations are considered from that overall population that differ in the distributions of the explanatory variables? In fact, the same regression model will hold then in each of these subpopulations. That is, correct regression models are invariant to this subpopulation choice. This is a particularly powerful feature of regression analysis as a statistical analysis and modeling methodology, which has contributed substantially to its popularity across different scientific disciplines (for example, Meredith [1993]).

The discussed invariance property is not shared, however, with widely used correlation coefficients in behavioral and social research. In particular, because the correlation coefficient depends in general on the group variability on both the predictors and outcome variables (treated symmetrically in the coefficient), as is well known, these coefficients change value in general when subpopulations are considered from an initial population. This is the widely known property of range restriction for correlation coefficients (for example, Agresti and Finlay [2009]). That is, the correlation coefficient is not invariant with respect to subpopulation choice.

Similarly, the popular number correct (NC) score, which is also referred to at times as "number right score" and defined for a prespecified set of binary items as the sum of the scores on them (sum of the 1s on them), depends on the characteristics of these items. In particular, if the items are "easy", all or most persons they are administered to will have high NC scores; conversely, if the items are "difficult", all or most persons will have low NC scores. That is, for a given group of persons, the properties of the NC score are entangled with those of the items, and vice versa.

What would be desirable instead is a procedure rendering i) such end scores for the studied subjects that are independent of the characteristics of the used items. In addition, we would want this procedure to yield ii) such indices for the items—specifically with respect to the way they function as individual measures, or their quality—that are independent of the characteristics of the persons to whom they have been actually administered. In this sense, we want a method for evaluation of i) subject ability or trait levels that are invariant with respect to the choice of items administered and ii) item characteristics that are invariant with respect to the choice of persons having taken them. In other words, the sought methodology should render individual trait or ability estimates (predictions) as well as item characteristic (parameter) estimates, which share the earlier mentioned invariance property of regression models.

We alluded to earlier in the book (see chapter 4 and section 5.1 of this chapter), and will see in further detail soon, that IRT models are in essence particular non-linear regression models. (Thereby, the so-called logistic IRT models, which are discussed in detail later in this chapter, are rendered in effect linear models via the logit transformation with respect to the response variables.) Hence, IRT models share the above indicated invariance property of regression models. More specifically, when an IRT model is correct in the population of interest, the resulting parameter estimates possess that invariance property, up to estimation or sampling error (for example, Hambleton, Swaminathan, and Rogers [1991]). Therefore, if such an IRT model is used, the parameter estimates resulting from it do not depend on a subpopulation of subjects or set of items used. This is due to the above invariance property of regression models. Based on it, the same ICCs follow regardless of the distribution of ability in the sample of persons used to estimate the item parameters. (This results from the fact that the value of the probability of "correct" response does not depend then on the number of persons located at a given ability level; compare Hambleton and Swaminathan [1985].) In other words (apart from estimation or sampling error) the trait, construct, or ability level estimates or predictions obtained when using such an IRT model do not depend on the choice of the items; conversely, the item parameter estimates do not depend on the group of persons (whose observed results are used to obtain the estimates).

This invariance feature holds in particular when a single trait underlies subject performance on a given set of items or measuring instrument. Then, using an IRT model that is correct in the population, one can place all examined persons and items on a common scale (continuum or dimension) even when some of the former have taken different subsets of an initially considered set of items (see van der Linden [2016b] and below for further qualification). This property is often referred to as test-free measurement with respect to subjects and group-free measurement with respect to item characteristics. That is, regardless which examined person has taken which (sub)set of items (instrument, test, or testlet), each person receives a particular placement on a common dimension, as does each item with respect to its so-called item difficulty parameter (see section 5.4). Thus, different subjects can be compared based on their resulting placement. That is, conclusions about persons can be made regardless of the actual item set administered to them (when an IRT model is used that is correct in a studied population). Similarly, conclusions about items can be made then even if different groups of individuals have taken them. The reason is in part the fact that particular item characteristics—the so-called item difficulty parameters—are evaluated as appropriate points on the same common scale across items and persons, so that these item comparisons become possible as well (see also below in this chapter).

While this invariance property is highly desirable for a measurement process, in particular in the behavioral, educational, and social sciences, we stress that for it to hold, the used IRT model must be correct, that is, perfect, in the studied population at large. One could of course argue that this is most of the time a requirement of theoretical relevance only that is not really empirically verifiable. The reason is that no realistic statistical model, and in particular no IRT model, can be in general perfect or completely correct. In addition, in empirical research, we typically have no way of knowing sufficiently well the performance of a given model in a studied population because as a matter of routine, the latter is not entirely accessible to a researcher. Nonetheless, this property of IRT models—which derives from the same property of regression models—is very appealing and a major reason why IRT is so popular in these and cognate disciplines.

With the preceding discussion in this chapter, we are now ready to take another look at the logistic function in preparation for our subsequent focus on specific logistic IRT models and demonstration of their applications using the widely circulated software Stata.

5.4 One more look at the logistic function

As was indicated in chapter 2, a possible and in fact widely followed choice for an ICC is as the curve described by the logistic function. This popular selection leads to logistic IRT models that are discussed in the remainder of the present chapter and used in the rest of the book. For completeness, we restate this function's formal definition next, using the earlier adopted notation $\Lambda(x)$ for it [the following equation is identical to (2.4) in chapter 2 and is presented here also for convenience $(-\infty < x < \infty)$]:

$$\Lambda(x) = \frac{e^x}{1+e^x} = \frac{1}{1+e^{-x}} = 1/\{1+\exp(-x)\} \tag{5.4}$$

As we mentioned before, the logistic function (curve) that underlies the widely used logistic IRT models is defined for all real numbers. This function has the following useful properties that are worth stressing next and are in fact directly seen from figure 1.1 in chapter 1 (see also figure 2.4 in chapter 2 and figure 5.2 below, which is identical to the former and displayed to make the present chapter self-contained):

- the curve described by the logistic function $\Lambda(x)$ rises continually (when x increases);

- its lower asymptote is 0 (that is, as x approaches $-\infty$, the curve gets closer and closer to 0), and its upper asymptote is 1 (as x approaches ∞); and

- it graphs, as a function of x, the area to the left of x and below the probability density function curve of the standard logistic distribution $(-\infty < u < \infty)$,

$$\psi(u) = \frac{e^u}{(1+e^u)^2} \tag{5.5}$$

whereby this area is (compare Roussas [1997])

$$A_\psi(x) = \int_{-\infty}^{x} \psi(u)du = \int_{-\infty}^{x} \frac{e^u}{(1+e^u)^2}du = \frac{e^x}{1+e^x} = \Lambda(x) \tag{5.6}$$

We recall from chapter 2 that the graphs of the logistic and normal ogive functions are indistinguishable for most practical purposes, after a minor rescaling of the horizontal axis for the former (with no substantive meaning in empirical settings in general). Thus, we can use figure 5.2 as a graphical representation of either of these functions for ICC illustration purposes in the rest of the book. (The units on the horizontal axis of the figure are not relevant for the present discussion.) In this figure, to prepare for the discussion in the remainder of the current and following chapters, we denote the horizontal axis by θ ("theta"), which is the underlying trait, construct, or ability of central interest in (unidimensional) IRT and its applications (see also chapter 1 and footnote 1 to it).

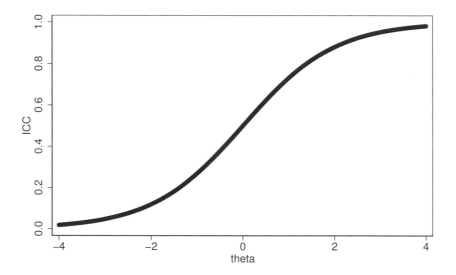

Figure 5.2. A typical (in shape) item characteristic curve in logistic item response theory models

We are now ready to discuss specific logistic IRT models that represent the over-whelming majority of IRT applications in contemporary behavioral, educational, and social research. These models are also used in a number of cognate disciplines ranging from biomedicine through marketing (for example, Raykov and Calantone [2014]).

5.5   The one- and two-parameter logistic models

When a logistic function is used as an ICC in a unidimensional IRT model, the height of the pertinent logistic curve at any given value of x, $\Lambda(x)$, obtains a special meaning, as it follows from (5.6) and is seen from figure 5.2. Specifically, this curve value $\Lambda(x)$ informs about the (assumed) proportion of persons in a population under investigation at that trait or ability level who can answer "correctly" the corresponding item. That is, from (5.5) and (5.6), it follows that the associated probability of correct response on the item is

$$P(\theta) = \int_{-\infty}^{x} \psi(u)du = \Lambda(x) \tag{5.7}$$

In (5.7), $\Lambda(x)$ denotes as before the standard logistic cumulative distribution function, and $\psi(u)$ is its corresponding probability density function [compare Roussas [1997] and see (5.5)]. Using the inverse of the logistic function (see chapter 4), (5.7) is interpretable as saying the following:

$$x = \Lambda^{-1}\{P(\theta)\} \tag{5.8}$$

We obtained (5.8) by taking the inverse function of both sides of (5.7) (namely, its first and last part), recalling thereby that the inverse of the logistic function exists because of the latter being monotonically increasing (see also chapter 2). We also kept in mind that inverse functions applied successively simply annihilate or wipe out each other, as indicated earlier.

Equation (5.8) actually states that x is a function of θ, as we can see by looking at its right-hand side. Indeed, because the right-hand side of that equation is a function of θ, namely, $\Lambda^{-1}\{P(\theta)\}$, its left-hand side is also a function of θ, that is, x is a function of θ as well. Hence, we can rewrite (5.8) now as follows to emphasize this dependence (see in particular the last part of the next equation):

$$x = \Lambda^{-1}\{P(\theta)\} = x(\theta) \tag{5.9}$$

We can interpret (5.9) as demonstrating the following important fact when considering the logistic function giving rise to the ICC of a binary or binary scored item of concern. Specifically, to obtain the point x at which the value of the ICC equals a given probability, say, P, we need to take the inverse of the function representing the cumulative distribution function of the standard logistic distribution, $\Lambda(\cdot)$, at that value P. That inverse function depends on the studied latent dimension θ, as we have just seen, and this is why we denoted it $x(\theta)$ in (5.9).

With this discussion in mind, we can now obtain one of the most popular logistic IRT models by taking one more "small" step as we do next.

5.5.1 The two-parameter logistic model

A widely used model in the behavioral, educational, and social sciences when there is no guessing (or only minimal such that can be treated as negligible) on any item in a given set or measuring instrument is based on (5.9). (For instance, such are items involving free response or items that are administered after effective instruction in an assessment setting.) This widely used model assumes that for a given binary or binary scored item, the dependence of x on the studied trait level θ as reflected in that equation is representable by the following linear relationship:

$$x = a(\theta - b) \tag{5.10}$$

In (5.10), a and b are parameters with special interpretation that are discussed in detail below. Because for a given item this model involves two parameters, it is referred to generically as a two-parameter model.

With (5.4) and (5.10) in mind, if we assume the logistic function as ICC for each item in a measuring instrument or item set of concern, from (5.7) follows that the formal representation of the ICC, that is, of the probability of "correct" response, is

$$P(\theta) = \Lambda(x) = \frac{e^x}{1 + e^x} = \frac{1}{1 + e^{-x}} = 1/\{1 + \exp(-x)\} \tag{5.11}$$

That is, owing to (5.10), from (5.11), we obtain

$$P(\theta) = \frac{e^{a(\theta-b)}}{1 + e^{a(\theta-b)}} = \frac{1}{1 + e^{-a(\theta-b)}} = \frac{1}{1 + \exp\{-a(\theta - b)\}} \tag{5.12}$$

An IRT model that has as associated ICCs the function in (5.12) for each in a given set of binary or binary scored items, with in general different values for its parameters a and b, is called a two-parameter logistic (2PL) model.

We stress that (5.12) defines an item-specific model. This model, when considered in an empirical setting, is usually assumed for any item in an instrument or item set of interest. That is, if the model is posited for, say, the jth item, in a set consisting of k items ($k > 1$), to be more informative, one should attach the subindex j to both a and b in the right-hand side of (5.12). This leads to the following widely used form of the 2PL model (for the jth item),

$$P_j(\theta) = \frac{e^{a_j(\theta-b_j)}}{1 + e^{a_j(\theta-b_j)}} = \frac{1}{1 + e^{-a_j(\theta-b_j)}} = \frac{1}{1 + \exp\{-a_j(\theta - b_j)\}} \tag{5.13}$$

$(j = 1, \ldots, k)$.

We will discuss in detail the meaning of the a and b parameters in the next subsection. Before doing so, however, we observe an important fact that follows from (5.12) [see also (5.13)]. For a fixed value of the quantity b (and θ), an increase in the quantity a leads to an increase in the probability $P(\theta)$ (for "correct" response). Conversely, a decrease in a then brings about a decrease in that probability. Similarly, for a fixed value of a (and θ), an increase in the quantity b leads to a decrease in the probability $P(\theta)$. Alternatively, a decrease in b then yields an increase in that probability.

Keeping in mind the 2PL model, we now note a consequential link between IRT and IRM on the one hand and our discussion in the preceding chapter on the other. More concretely, through simple algebra on (5.13), we obtain

$$P_j(\theta) = \frac{e^{a_j(\theta-b_j)}}{1 + e^{a_j(\theta-b_j)}} = \frac{1}{1 + \exp\{-a_j(\theta - b_j)\}} = \frac{1}{1 + \exp\{-(c_j + a_j\theta)\}} \quad (5.14)$$

for the jth item, where $c_j = -a_jb_j (j = 1, \ldots, k)$. Comparing now each of the k equations in (5.14) with the corresponding equations (for the same items) in (4.12) in chapter 4 defining the multivariate logistic regression model with a single observed predictor, we can make the following observation. We can view the 2PL model, and by implication the Rasch model as a special case of it (see further details below), as a multivariate logistic regression model with i) a single unobserved predictor, namely, θ; ii) intercept $c_j = -a_jb_j$; and iii) slope a_j for the jth item in a set of items or measuring instrument of concern ($j = 1, \ldots, k$; compare Cai and Thissen [2015]). We will provide an alternative yet equivalent view of this relationship later in the chapter, after attending next to the particular meaning of the item parameters in this model.

5.5.2   Interpretation of the item parameters in the two-parameter logistic model

What do these parameters a_j and b_j actually mean in the 2PL model in relation to the jth item ($j = 1, \ldots, k$)? The parameter a_j can be shown to be directly proportional to the steepness of the ICC at its inflection point (for example, Reckase [2009]). In particular, the higher a_j, the steeper the slope of the ICC in its central part, and conversely (see figure 5.2). This inflection point, as mentioned in chapter 2, is located at that trait or ability level (point on the horizontal axis in an ICC graph) where the probability of correct response is 0.5. (See, for example, figure 2.2 in chapter 2 or figure 5.2 above. We mention in passing that this trait or ability level or point on the θ-scale is at times also referred to as the "midprobability" point for the considered item.) Looking at (5.12), we readily realize that this happens precisely where $\theta = b_j$ on the underlying continuum representing the studied trait or ability. (The reason is that only at this point, the numerator of the ratio in the right-hand side of this equation is equal to 1 and its denominator is equal to 2.) Thus, the meaning of the parameter b_j is as that position on the latent ability or construct scale (dimension or continuum), where $P_j(\theta) = 0.5$ holds for the probability of correct response on the jth item (see below for additional discussion).

This interpretation of the b parameter for an ICC (item) in a 2PL model also shows a more general and highly useful feature of IRT models. Specifically, through this parameter, items can actually be also thought of as being "positioned" on the same latent dimension on which persons are in terms of their individual trait or levels, denoted $\theta_1, \ldots, \theta_n$. (As will be recalled, n is used in this book to denote sample size.) That is, as a result of an application of IRT, we obtain estimates of the individual trait or ability levels of all persons, $\theta_1, \ldots, \theta_n$, as well as of the "locations" of each of the items, namely, their b parameters, that is, $b_1, \ldots, b_k$. In other words, we achieve "measurement" of both persons and items, assuming of course that a plausible IRT model is used for this purpose, in terms of the same underlying dimension, θ. The relationships among these $n + k$ quantities (estimates), $\theta_1, \ldots, \theta_n$ and $b_1, \ldots, b_k$, depend on the data and obviously on the studied persons and used items, given a plausible IRT model. In this connection, we should keep in mind that at the end of an IRT application, these $n+k$ resulting quantities can be treated as located along the same single underlying dimension, in unidimensional IRT (see also van der Linden [2016b]; compare chapter 12).

   The a and b parameters for an item in a measuring instrument or item set subjected
to IRT analysis using a 2PL model that is plausible for an analyzed study dataset are of
major importance in theoretical and empirical research using IRT modeling. Consistent
with their above interpretation, these two parameters have received special names. In
particular, for a given item, its a parameter is called an "item discrimination parameter"
and its b parameter is referred to as an "item difficulty parameter". The reason for the
former name is that the steepness (tilt) of an ICC in its central part, which as mentioned
above is directly proportional to the a parameter (Baker and Kim 2004; Reckase 2009),
has to do with the capability of that item to differentiate well between persons with
trait or ability values that are close to its b parameter (location) on the θ-scale and
possibly on different sides on that location. Specifically, the higher the a parameter,
the more pronounced the difference in their probabilities for "correct" answer (see, for
example, figure 5.2).[1] Similarly, the larger the b parameter, the higher the level of the
underlying trait or ability that is required to obtain probability of 0.5 for correct

---

1. As a simple example illustrating this discussion, one can readily use Stata to compute under
the 2PL model the difference in the probabilities of correct response for two subjects that are,
respectively, say, 0.5 units (standard deviations) above and below the mean on a latent dimension,
using two items that differ only in their discrimination power. (Recall that the mean on the latent
dimension is set at 0 and its variance at 1.) For the sake of this example, consider item 1 as
having discrimination parameter $a = 1$, say, and item 2 with discrimination parameter $a = 1.5$.
To keep the setting underlying figure 5.3 that fixes their difficulty parameters, assume finally for
convenience that for each of these items, $b = 0$. Then let us use the pertinent 2PL model (5.12)
for the probability of "correct" response; that is,

$$P_j(\theta_i) = \exp\{a_j(\theta_i - b_j)\}/[1 + \exp\{a_j(\theta_i - b_j)\}] \tag{5.15}$$

($j = 1, 2; i = 1, 2$). Using (5.15), for item 1, we obtain with Stata the difference in these probabilities for the two subjects in question as follows (resulting probability difference presented beneath
command):

```
. display exp(1*(.5-0))/(1+exp(1*(.5-0))) - exp(1*((-.5)-0))/(1+exp(1*(-.5)-0))
.24491866
```

At the same time, for item 2, this probability difference, for the same subjects, is

```
. display exp(1.5*(.5-0))/(1+exp(1.5*(.5-0))) -
> exp(1.5*((-.5)-0))/(1+exp(1.5*(-.5)-0))
.3583574
```

Hence, an increase in the item discrimination parameter by a half is associated here with a considerable increase in the difference in the (same) subjects' probabilities of correct response, that
is, leads to markedly enhanced subject differentiability. Note that this increase in item discrimination parameter renders these subjects' correct response probabilities to differ by almost 50%
as well in this example (relative to the item possessing a lower discrimination parameter), which
is a substantial discrepancy in the difference in probability of correct response across the items.
(Observe that here $a_1 = 1$, $a_2 = 1.5$, $b_1 = b_2 = 0$, $\theta_1 = 0.5$, and $\theta_2 = -0.5$.)

response on the item, that is, the more difficult the item is, and vice versa (see also figure 5.2).[2]

To illustrate graphically this feature of the item discrimination or a parameters, figure 5.3 displays the ICCs of several items that differ from each other only in these parameters. [See chapter 2 on how one can generate with Stata any of these curves implementing (5.10) and (5.13), as well as those on figure 5.4 below.][3] In figure 5.3, the gradient (or tilt) of the ICC in its "central" part becomes steeper and steeper with increasing item discrimination parameter. This increment is also to be expected because the ICC under the 2PL model is an increasing function in the item discrimination parameter a, as noticed from (5.12) and mentioned earlier (for given θ and b). We stress that all ICCs in figure 5.3 have the same b parameter because they intersect at the same point at which the corresponding probability of "correct" response is 0.5 for each item.

---

2. An extended interpretation of the difficulty parameter for a given item is directly obtained with this discussion in mind. (This interpretation will turn out to be rather useful in chapter 11 on polytomous items and IRT models, where the item responses will be more generally referred to as "response categories".) Specifically, viewing the scores 0 and 1 on a dichotomous item of concern in this chapter as such associated with "choosing" the corresponding from the two possible response categories on it, its b parameter can be seen as the propensity to choose response category "1" versus response category "0", for a given value of θ. If formally considering the latter response as a "baseline category", this interpretation describes the item difficulty parameter as the propensity to choose a "higher" category relative to the baseline category (in this chapter, as the propensity to choose the category scored "1" instead of the category scored "0"). In particular, the lower the b is, the stronger the propensity to choose the category "1" instead of the baseline category (response scored "0"), at a given trait or ability level, and vice versa (see also chapter 11).

3. For instance, to generate a plot with the ICCs of 4 items with $b = 0$ and a parameters being 1, 2, 3, and 4, use this Stata command (compare figure 5.3):

```
. twoway function logistic(x), range(-4 4)
> ||function logistic(2*x), range(-4 4)
> ||function logistic(3*x), range(-4 4)
> ||function logistic(4*x), range(-4 4)
```

Similarly, to generate a plot with the ICCs of 4 items with $a = 1$ and b parameters being 0, 1, 2, and 3, use this Stata command (compare figure 5.4):

```
. twoway function logistic(x), range(-4 4)
> ||function logistic(x-1), range(-4 4)
> ||function logistic(x-2), range(-4 4)
> ||function logistic(x-3), range(-4 4)
```

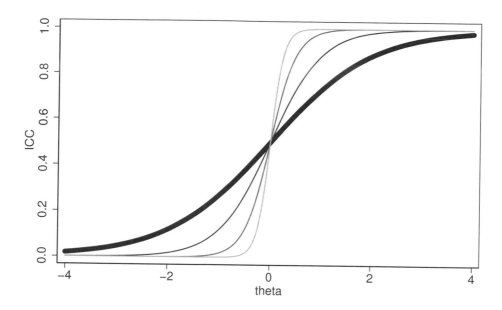

Figure 5.3. Illustration of increasing steepness (in their central part) of item characteristic curves for items with increasing item discrimination parameter and same difficulty
parameter

The above interpretation of the b parameters is illustrated in figure 5.4. The figure displays the ICCs of several items differing from each other only in their difficulty
parameter. In particular, the entire ICC is moved or shifted to the right as the items
become more difficult (that is, are associated with increasing b parameters). This is
also to be expected because the ICC under the 2PL model is a decreasing function in the
item difficulty parameter b, as noticed from (5.12) (for fixed a and θ).

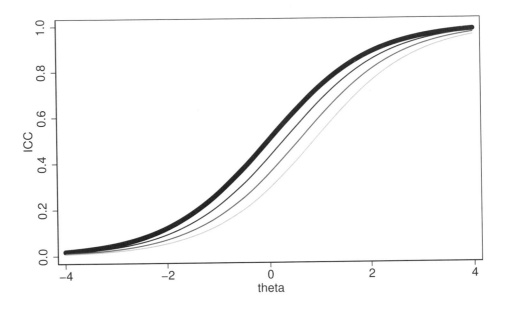

Figure 5.4. Illustration of effect of increasing item difficulty parameter (namely, "shift" of the item characteristic curve to the right) for a fixed item discrimination parameter

From this discussion, we readily see that items with nearly flat ICC's are not differentiating well—if at all, in practical terms—between persons with ability levels around (close to) their b parameter. (See, for instance, the item with the thickest ICC in figure 5.3, which is associated with the smallest discrimination parameter, and footnote 1 for an additional example.) These are items with relatively small a parameters. The items therefore possess the property that the difference in the probability of correct response is small when comparing persons with trait levels relatively close to each other and in the central part of the ICC, for instance, around its inflection point [recall from (5.12) that for given θ and b, $P(\theta)$ is an increasing function of a; see also footnote 1]. Hence, an index of quality of an item when the aim is to differentiate between persons (as is usually the case in behavioral and social measurement) is the steepness of the ICC, that is, the value of the a parameter. Specifically, the larger the value of a, the better the item, in particular for the purpose of differentiating among subjects close to its difficulty parameter.[4] Thus, for a set of items with the same b parameter but different a parameters, if the goal is to choose the items differentiating best among persons, we could consider selecting the items with the highest a parameters. (We will provide further qualification of this recommendation in chapter 9, where particular ranges of the underlying θ-scale will be of focal interest.)

Similarly, notice that in addition to items with small a, also items with $a < 0$ are not desirable for the same purposes (in a set of items where all others have positive discrimination parameters). This is because such items have a counterproductive contribution to the overall instrument functioning, because they suggest that increasing trait or ability level leads to lower probability of "correct" response, other things being the same (a possibility to keep in mind for such items is of course that they may be associated with reverse, or potentially erroneous, coding of "correct" and "incorrect" responses). With these observations in mind, in the rest of the book, we will assume that the discrimination parameter a is positive, that is, $a > 0$, for any considered item.

5.5.3 The scale of measurement

As we have repeatedly stated in the book, the latent trait or ability of interest, denoted θ, which is assumed to underlie studied subjects' performance on a set of items, cannot be directly measured because it is not observable. Hence, its "scale" does not have a

---

4. As can be seen from this discussion, items with small a parameters do not differentiate well between subjects whose trait or ability levels are close to their b parameters. Moreover, items with nearly flat ICCs in their "central" part (that is, around their inflection or midprobability point) in general do not differentiate well, if at all in practice, between persons with ability levels positioned just about anywhere on the underlying θ-scale. Indeed, using the ICCs of such items as functions of these ability levels (see, for example, figure 5.1), we readily observe very small if at all meaningful differences in the associated probabilities of "correct" response for persons with θ values close to these items' difficulty parameters, or elsewhere on the θ-scale. Conversely, we notice easily also that while items with large a parameters differentiate well between subjects with trait or ability levels close to their b parameters, they do not differentiate well if at all in practical terms among subjects with trait levels located (sufficiently) further away from their b parameters in either direction. This observation will become particularly relevant in chapter 9, which deals with measuring instrument construction and development.

natural origin and unit of measurement. In fact, there is no way to come up with such entities unless they are chosen (fixed) in an essentially arbitrary fashion. (Recall the similar circumstance in factor analysis, where we in effect arbitrarily assumed the factors to have a mean of 0 and variance of 1 for the same reason; see chapter 3.) Therefore, in IRT, and specifically IRM, it is common to select (fix) the origin and unit on the latent scale, that is, the θ-scale, so that the mean latent trait is 0 and its standard deviation is 1 for a population of interest. Hence, on any graph of an ICC within an IRT model, indication of negative scores on the horizontal axis (representing the latent trait θ) should not be viewed as unusual. In fact, they are in general just as usual or common as positive scores on that axis (see also footnote 1 to chapter 1).

Furthermore, we easily see that the (choice of the) origin and unit of the latent trait scale do affect the values of the a and b parameters for a given item but not the probability of "correct" response. Indeed, the following relationship is directly seen from (5.10) and in particular (5.13). Increasing, for instance, the b parameter by a certain amount (number) can be compensated by decreasing with the same amount the θ value, and conversely (other things being the same), leaving in the end unchanged the x value in (5.10) that determines this probability within the 2PL model. That is, these changes in b and θ do not alter the probability of "correct" response [see (5.13)]. Similarly, increasing the a parameter by any amount (that is, say, by multiplication with a number) can be compensated by division of the θ and b values by that same amount (number), leaving again unaltered the critical x value and that probability [see (5.10) and (5.13)]. Thus, owing to the probability $P_j(\theta)$ of "correct" response on the jth item depending only on the product $a_j(\theta - b_j)$, this probability is not affected by the selected origin or unit of the latent scale ($j = 1, \ldots, k$). Therefore, the probability of correct response is unrelated to the choice of origin and unit on the latent scale.

Thus, we can conclude that the probability of correct response on a given item, according to a 2PL model, is invariant under changes in origin and location of the underlying latent trait dimension. (This probability invariance is in fact the case for any IRT model mentioned in the present chapter or later in the book.) The cause of this invariance is in effect a phenomenon referred to as "lack of identification" in applied statistics. In fact, unless this identification issue is properly dealt with, we cannot obtain unique estimates of the a and b parameters associated with any item, given the data on a set of k items under consideration ($k > 1$). To resolve this problem, that is, to identify the model parameters, we will assume in the rest of this book (unless otherwise stated) that the mean of the individual trait or ability levels is 0 in the studied population and its variance is 1, as is usually done in IRT applications. We note that this identification problem also arises in other areas of applied statistics and is closely related to that in factor analysis and structural equation modeling (for example, Raykov and Marcoulides [2006, 2008]).

5.5.4   The one-parameter logistic model

A special case of the 2PL model is the one-parameter logistic (1PL) model. This model is at times also called the Rasch model, particularly in contexts concerned with model fit evaluation. More concretely, the 1PL model is obtained from the 2PL model when it is in addition assumed that all items in an instrument or set of interest have the same discrimination parameter; that is,

$$a_j = a_r \qquad (5.16)$$

holds for all $j \neq r (1 \leq j \neq r \leq k)$. Hence, in the 1PL model, all items share the same gradient (steepness or tilt) of the ICCs at their inflection points. That is, the ICCs of all items are equally steep in their central part and thus can be seen as "parallel" to each other there. However, the items do differ in general in the positioning of their ICCs along the latent trait dimension; that is, they differ in their difficulty parameters,

$$b_j \neq b_r$$

for $j \neq r (1 \leq j, r \leq k)$. Thus, different items need not have the same difficulties (location parameters) in the 1PL model (Rasch model). This implies that some items are positioned "further to the right", that is, are more difficult, while others further "to the left"—that is, are less difficult—on the continuum representing the underlying latent trait, θ.

We stress that the 1PL model is in fact a 2PL model in which the relationships or restrictions in (5.16) hold. That is, the 1PL model is a restricted 2PL model. In general terms, models that are advanced for the same set of observed variables (dataset) but differ in complexity are called nested if one of them is obtained from the other by imposing one or more constraints on its parameters. With the preceding discussion in mind, we see that the 1PL model is nested in the 2PL model. This is because the 2PL model assumes in general different item discrimination parameters, while the 1PL model has only one parameter for all of them. The latter is a consequence of stipulating these parameters being equal across items [see (5.14) and (5.16)]. Stated differently, we can consider the 2PL model is a "full" model or more complex model with more parameters, whereas the 1PL model as a "reduced" or restricted model. We stress that the 1PL model is more parsimonious than the 2PL model because the former has $k-1$ fewer parameters than the latter.

5.5.5   The one-parameter logistic and two-parameter logistic models as nonlinear factor analysis models, generalized linear models, and logistic regression models

As stated earlier in the book, we will be concerned with the setting of binary or binary scored items until chapter 11. For this setting, we can readily rewrite (5.12) by taking the logit of the probability $P(\theta)$ appearing in its left-hand side. This leads to the following alternative representation of the ICC for an item according to the 2PL model [see also (4.9)]:

$$\text{logit}\{P(\theta)\} = a(\theta - b) = (-ab) + a\theta \qquad (5.17)$$

However, we are usually dealing in empirical utilization of IRT with a set of items comprising a measuring instrument of concern, which consists of, say, k items ($k > 1$). Hence, stating the 2PL model in (5.17) for each item yields the following set of nonlinear equations, because their left-hand sides are nonlinear functions of the probability of "correct" response on the pertinent item:

$$\text{logit}\{P_1(\theta)\} = a_1(\theta - b_1) = (-a_1 b_1) + a_1 \theta$$
$$\text{logit}\{P_2(\theta)\} = a_2(\theta - b_2) = (-a_2 b_2) + a_2 \theta$$

$$\vdots$$

$$\text{logit}\{P_k(\theta)\} = a_k(\theta - b_k) = (-a_k b_k) + a_k \theta \qquad (5.18)$$

We can represent the system of (5.18) more compactly as

$$\text{logit}\{P_j(\theta)\} = a_j(\theta - b_j) = (-a_j b_j) + a_j \theta = c_j + a_j \theta$$

where $c_j = -a_j b_j$ is used as simplifying notation; that is, $b_j = -c_j/a_j (j = 1, \ldots, k)$. (Recall our previous assumption that the discrimination parameter does not vanish for any considered item, which we made for the remainder of the discussion in this book.) This representation of (5.18), as well as its associated expressions for the probabilities of "correct" response on the items, is sometimes referred to as the "intercept-and-slope" form of the 2PL model in the literature, and similarly used more generally with logistic models (for example, Cai and Thissen [2015]).

With these considerations in mind, let us now compare (5.18) with (4.17) in chapter 4, which represented the nonlinear factor analysis model with the logit link. Based on this comparison, we observe that (5.18) can actually be seen as describing a nonlinear factor analysis model with $m = 1$ trait, namely, θ [using formally in (5.18) the notation $c_j = \mu_j$, $\theta = f_1$, and the a's as pertinent factor loadings]. From this vantage point, the 2PL model is a nonlinear factor analysis model with a single factor. Hence, in this sense, the 1PL model (Rasch model), being a special case of a 2PL model, can also be viewed as a nonlinear factor analysis model with a single factor. In fact, the 1PL model is seen then as a restricted nonlinear factor analysis model with equal loadings of that common factor [see (5.16) in the preceding subsection].

We next recall that in the present binary item setting, the mean (expectation) of the RV defined as response on an item, Y, equals the probability of "correct" answer on it (for example, chapter 3). Hence, we can also reexpress the above (5.17) as follows,

$$\text{logit}\{\mu_Y(\theta)\} = c + a\theta \qquad (5.19)$$

where $\mu_Y(\theta)$ designates the mean of the dependent variable, Y, at a given trait level, θ, and $c = -ab$.

Equation (5.19) states that a particular function of the mean of the response variable Y, rather than the mean of Y, equals a linear function of the underlying latent trait, θ. This function represents what can be seen as a linear predictor here, namely, a linear function of the unobserved predictor (see chapter 4). Hence, we can view conceptually the 2PL model also as a generalized linear model (GLIM) with a single unobserved predictor θ (linear predictor, or systematic part), Bernoulli random component, and the logit function as a link function (for example, De Boeck and Wilson [2015]; Mellenbergh [1994]; Rabe-Hesketh and Skrondal [2016]; Muthén and Asparouhov [2016]).

As we explained in chapter 4, the widely used (binary) logistic regression model is a special case of the GLIM. With this fact and the preceding discussion in mind, it follows that (5.19) also defines a binary logistic regression model for the dichotomous item response, Y, with a single predictor that is not observed, θ. In particular, if we write (5.19) for each of the k items of concern here, we will obtain the following set of logistic models across the items,

$$\text{logit}\{\mu_{Y_1}(\theta)\} = \text{logit}\{P_1(\theta)\} = c_1 + a_1\theta$$
$$\text{logit}\{\mu_{Y_2}(\theta)\} = \text{logit}\{P_2(\theta)\} = c_2 + a_2\theta$$
$$\vdots$$
$$\text{logit}\{\mu_{Y_k}(\theta)\} = \text{logit}\{P_k(\theta)\} = c_k + a_k\theta \qquad (5.20)$$

where $Y_1, \ldots, Y_k$ denote the responses on the respective items. Based on these considerations, and as an alternative view on the earlier (5.14), (5.20) can now be seen as a multivariate binary logistic model—and thus a multivariate nonlinear regression model—with i) a single unobserved predictor, θ; ii) intercepts c_j being the negatives of the products of the item difficulty and discrimination parameters (that is, $c_j = -a_j b_j$); and iii) slopes being the item discrimination parameters a_j ($j = 1, \ldots, k$; compare Cai and Thissen [2015]; see also chapter 4).

5.5.6 Important and useful properties of the Rasch model

The popular Rasch model, named after the Danish statistician Georg Rasch, is statistically equivalent (in terms of model fit) to the earlier discussed 1PL model. That is, the Rasch model, as a 1PL model, is nested in the 2PL model and obtained from it by imposing the parameter constraint of equal discrimination parameters across the k items of concern:

$$a_1 = a_2 = \ldots = a_k = a \text{ say} \qquad (5.21)$$

We note that in some alternative parameterizations of the Rasch model, the common item discrimination parameter a can be assumed to be 1. It is important to keep in mind, then, that in this model reparameterization, the latent variance is a free parameter.

From our earlier discussion in section 5.5, it follows that the Rasch model can be viewed as i) a nonlinear factor analysis model with a logit link and a single trait (factor), θ; ii) a GLIM with Bernoulli response, logit link, and a linear predictor in that unobserved ability or construct; and iii) a multivariate binary logistic model, as well as a multivariate nonlinear regression model, with intercepts being the item difficulty parameters when

$a = 1$ and an item-invariant slope as the common item discrimination parameter a, that is, 1 if $a = 1$ is assumed [see (5.21) and preceding paragraph].

When the Rasch model is plausible for a given dataset (studied population), all of its items have the same discrimination power. Because of this identity, unlike the case with the 2PL model, the ICCs for a set of items following the Rasch model are all parallel in their central parts. These ICCs can thus be thought of as being obtained from one another via shifting to the right or to the left along the horizontal axis (representing the trait or ability studied, θ) by as many units as are the respective differences in their difficulty parameters.

From (5.18) and (5.21), it follows that for a given item (say, the jth), the Rasch model postulates the logit of the probability of "correct" answer as

$$\text{logit}\{P_j(\theta)\} = a(\theta - b_j) \tag{5.22}$$

($j = 1, \ldots, k$). Let us now take the inverse of the logit, namely, the logistic function, from both sides of (5.22). This will show that in the Rasch model, the probability of obtaining correct response on a given item is represented as follows [note the lack of subindex attached to the discrimination parameter a, which is due to the imposed constraints in (5.21)],

$$P_j(\theta) = \frac{e^{a(\theta - b_j)}}{1 + e^{a(\theta - b_j)}} = \frac{1}{1 + e^{-a(\theta - b_j)}} = \frac{1}{1 + \exp\{-a(\theta - b_j)\}} \tag{5.23}$$

($j = 1, \ldots, k$).

Part of the particular attraction of the Rasch model when plausible for an analyzed dataset (population under investigation), in addition to its parsimony, lies in the fact that this model possesses an important property allowing an insightful interpretation i) of the differences in the abilities of two persons taking the same item and ii) in the difficulties of two items administered to the same person. This property is closely related to the concept of odds for "correct" response on an item, which we denoted w earlier in this book (see chapter 2).

To highlight this property of the Rasch model, let us assume we are interested in the response of a subject with trait level θ on a given binary or binary scored item (compare Hambleton and Swaminathan [1985]). As discussed in chapter 2, the odds for his or her "correct" response are

$$w(\theta) = P(\theta)/\{1 - P(\theta)\} \tag{5.24}$$

where $P(\theta)$ is the probability for this response (and for convenience, we drop the item subindex j used above). That is, in addition to the associated true score and this response probability, the odds for "success" on the item depend on the underlying trait or ability level, θ. Assuming the Rasch model [see (5.22) and (5.23) for a given item],

$$P(\theta) = \exp(\theta - b)/\{1 + \exp(\theta - b)\} \tag{5.25}$$

holds for this probability, with the parameterization setting $a = 1$, which is also used for simplicity and convenience throughout the rest of this subsection. (In this parameterization, we reiterate that the latent variance, that is, the variance of the latent trait θ in the studied population, no longer needs to be fixed at 1 as earlier but would instead be considered a free parameter; this fact is irrelevant, however, for the presently developed argument.) Substituting (5.25) into (5.24), after some algebra, we obtain

$$w(\theta) = \exp(\theta - b) = e^{\theta - b} = e^{\theta}/e^{b} = \theta'/b' \quad \text{say} \tag{5.26}$$

where the last two symbols, θ' and b', are used for simplicity to denote e^{θ} and e^{b}, respectively.

Suppose now that two persons with trait levels θ_1 and θ_2 have taken the same item, which has a difficulty parameter b, say. Then, based on (5.26), their odds for correct response, or "success", on the item are as follows (compare Hambleton, Swaminathan, and Rogers [1991]):

$$w(\theta_1) = \theta_1'/b' = w_1 \quad \text{say, and}$$
$$w(\theta_2) = \theta_2'/b' = w_2 \quad \text{say}$$

Therefore, the ratio of their odds for success on the item, or the odds ratio, is

$$w_1/w_2 = \theta_1'/\theta_2' \tag{5.27}$$

From (5.27), it follows that one may see the ability scale, under the Rasch model, as having the properties of a ratio scale in the following sense (for example, Hambleton and Swaminathan [1985]). If person A has twice the ability of person B, say, as measured on the scale resulting from the nonlinear transformation e^{θ} of the original θ-scale, then A has twice larger odds of success (correct response) on the considered item than B. In addition, one can similarly show that if item 1 is twice as hard as item 2, as measured on the scale resulting from the transformation e^{b}, then under the 1PL model, a given person has twice the odds for success on item 2 relative to his or her odds for success on item 1.

The preceding discussion in this subsection allows us also to observe a related property of the Rasch model (compare Hambleton, Swaminathan, and Rogers [1991]). Taking the logarithm of both sides of (5.27) and using (5.26) (with pertinent notation), we obtain

$$\ln(w_1/w_2) = \theta_1 - \theta_2 \tag{5.28}$$

for the case of two persons being administered the same item who have correspondingly trait or ability levels θ_1 and θ_2.

Equation (5.28) now allows the following useful interpretations. A difference of 1 point on the scale of the studied trait, construct, or ability (that is, on the θ-scale) corresponds to a factor of 2.72 (rounded off) in the odds of success or "correct" response on the item. In other words, an increase in ability by 1 unit on the ability scale is associated with odds of correct response on a given item that is multiplied by 2.72 relative to those odds at the initial position on that scale. Conversely, a decrease in ability by 1 unit on its scale leads to odds of success on the item being divided by 2.72.

2

2

222222222222

333

Similarly, for a given person taking 2 items, it can be shown that a difference in 1 unit in item difficulties (which are measured on the same scale as ability, as pointed out earlier) corresponds to a factor of 2.72 in the odds of success on the item. In other words, increasing item difficulty by 1 unit on the ability scale leads to division by 2.72 of the odds of success of that person on the item; conversely, decreasing item difficulty by 1 unit on the ability scale leads to multiplication by 2.72 of the odds of his or her success on the item.

Based on these interpretations, it becomes meaningful to use the reference "logits" for the units on the ability scale (log-odds scale, or θ-scale) when the Rasch model is plausible for a studied population. Because the logit of the probability of correct response is under the Rasch model the difference between ability and item difficulty, namely,

$$\text{logit}\{P(\theta)\} = \theta - b \tag{5.29}$$

we can actually think or talk of subjects' ability and item difficulty as being "measured" (evaluated, assessed, estimated) on a scale where the units of measurement are logits. We can thus refer to it as a "logit scale".

We similarly stress that the following important observation is readily made from (5.29). An increase by 1 unit (that is, 1 logit) in ability could be "compensated" for by a decrease in item difficulty by 1 logit, so that the probability of correct response on a given item remains unchanged, according to the Rasch model. Alternatively, increase or decrease in ability by any number of logits (units on that ability scale), while keeping item difficulty constant, leads to higher or lower probability of correct response (under the Rasch model). Conversely, increase by any number of logits in item difficulty while keeping ability the same leads to lower probability of correct response, and vice versa. These properties of the Rasch model follow directly from the fact that as a function of probability, the logit is increasing in the latter; conversely, the probability for correct response is an increasing function of the logit (see chapter 2).

The discussion in this subsection on the Rasch model has demonstrated several rather insightful interpretations of main IRT concepts within this model. These and related interpretations, in addition to the parsimony of the Rasch model, have contributed substantially to the popularity of the Rasch model in behavioral and social research. These features enhance the Rasch model's relevance in empirical settings, where it is a plausible means of description of datasets stemming from studied sets of items and samples from subject populations under investigation.

5.6 The three-parameter logistic model

While the 1PL model is a special case of the 2PL model, an alternative generalization of the 2PL model is applicable in behavioral measurement settings when guessing is possible on items in a given set or measuring instrument of concern. In such contexts, a person not possessing to a sufficient degree the necessary knowledge, ability, or trait level allowing him or her to find the correct answer on an item may actually "guess" that answer, or more generally, still get the correct response by chance. For instance, this is possible, as is well known, in tests consisting of multiple-choice items. In those and other related cases, a generalization of the 2PL model incorporates an added item parameter. The latter represents the probability of correct response by chance on an item under consideration, for instance, by subjects with very low values of the studied trait, ability, or construct.

This probability of chance-based success on an item, that is, selecting the correct response based on a chance-driven decision, is also unknown, typically considered an item parameter, and referred to as the pseudoguessing parameter for that item. This parameter is denoted as g throughout the present book ($0 < g < 1$). (We mention that in some treatments of the topic, the pseudoguessing parameter is symbolized by the letter c to alphabetically reflect that it is the third parameter in the model for an item. However, we prefer using the letter g for it because this better communicates the characteristic feature of this presumed item parameter.) The pseudoguessing parameter is involved in an essential way in an extended 2PL model, which is therefore called a three-parameter logistic (3PL) model (compare van der Linden [2016b]).[5]

According to the 3PL model, the probability of correct response on the jth item (allowing for guessing) is

$$P_j(\theta) = g_j + (1 - g_j)\frac{e^{a_j(\theta-b_j)}}{1 + e^{a_j(\theta-b_j)}} \qquad (5.30)$$

($j = 1, \ldots, k$; compare Raykov and Marcoulides [2011]). We note that the 3PL model is actually not a logistic model, strictly speaking, because the right-hand side of (5.30) cannot be reexpressed in a form subsumed under the logistic function [compare (5.11) through (5.14)]. However, because the reference to this model as a 3PL model has been so widely adopted and followed, we use it in the present book as well. (More important logically than a particular reference is obviously the defining equation of an IRT model; hence, as long as its definition (5.30) is kept in mind, the reference "3PL model" to this model cannot be really misleading.)

An example ICC curve for an item following the 3PL model, with the value of the pseudoguessing parameter being $g = 0.2$, is given in figure 5.5. We can readily obtain the ICC for an item according to the 3PL model using the following Stata command, where for the sake of this example, we have taken its discrimination parameter to be $a = 1.5$ and its difficulty parameter to be $b = -1$:

```
. twoway function ICC_3PL=.2+.8*exp(1.5*(x-(-1)))/(1+exp(1.5*(x-(-1)))),
> range(-4 4)
```

This command yields the graph displayed in figure 5.5 (the notation on the horizontal axis is similarly assigned as default by the software and, like earlier in the book, represents the underlying latent dimension θ).

---

5. A three-parameter normal ogive model is also possible to use in such circumstances. We will not be concerned with it in this book, however, which does not consider in more detail normal ogive models because of mathematical tractability and numerical approximation reasons mentioned earlier (see, for example, Hambleton and Swaminathan [1985]; see also chapter 2).

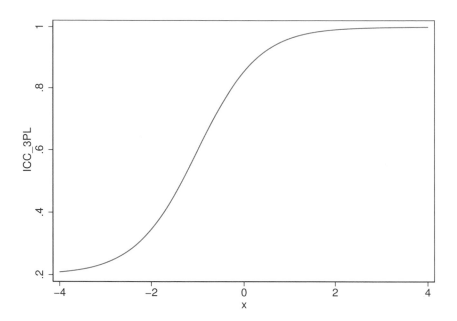

Figure 5.5. Item characteristic curve for an item following a three-parameter logistic model with pseudoguessing parameter $g = 0.2$ (and discrimination and difficulty parameters $a = 1.5$ and $b = -1$, respectively)

We note from figure 5.5 a generally applicable observation for 3PL models. This is the fact that unlike the ICCs for items following any of the models discussed so far in the book, the lower asymptote of the ICC within the 3PL model is the pseudoguessing parameter g—for example, 0.2 in figure 5.5—rather than 0 (compare section 5.4 and subsequent discussion of the 2PL model).

5.7   The logistic models as latent variable models and analogs to nonlinear regression models

5.7.1   Item response models as latent variable models

The preceding discussion in this section defined three main types of logistic models: the 1PL, 2PL, and 3PL models. Although they differ in consequential ways from one another, and specifically in the number of parameters associated with each of a given set of items, they all share one very important feature: they involve a latent variable, namely, the unobserved trait, ability, dimension, or continuum θ of actual interest (see below).

More generally, the essence of any IRT model used or referred to throughout this book is that it relates one or more latent variables—a single θ in the unidimensional IRT case and a set or vector of θ's in the multidimensional IRT case (see chapter 12)—to the items or measuring instrument of concern. With this perspective, any IRT model is in fact a latent variable model because it instrumentally involves at least one latent variable (denoted generically θ). The latent variable or variables are presumably manifested in the items used to evaluate it. We will keep in mind this important view throughout the rest of the book.

5.7.2 The logistic models as analogs to nonlinear regression models

Earlier in this chapter, we alluded to the fact that IRT models expressing the probability of "correct" response on binary or binary scored items can be viewed as nonlinear regression models. Indeed, looking at the defining equations for the 1PL, 2PL, and 3PL models in the preceding discussion [see (5.12) for the 1PL and 2PL models and (5.30) for the 3PL model], we see in their left-hand side the probability of correct response while on the right is a nonlinear function of the discrimination and difficulty parameters and the individual trait, construct, or ability levels, θ. Because in the context of binary items (of interest to us until chapter 11), this probability is also the mean of the response variable, denoted Y as before, we can rewrite these two main model equations as follows (with subindices indicating also model class, and dropping the item subindex j for simplicity):

$$\mu_{Y,2\mathrm{PL}}(\theta) = \frac{e^{a(\theta-b)}}{1 + e^{a(\theta-b)}}$$

$$\mu_{Y,3\mathrm{PL}}(\theta) = g + (1-g)\frac{e^{a(\theta-b)}}{1 + e^{a(\theta-b)}} \tag{5.31}$$

In (5.31), we stress that in their left sides is the mean of the response Y (as a function of θ), while in their right side is an assumed nonlinear function of item parameters and individual ability or trait levels. Specifically, in the first equation for the 2PL model (and correspondingly for the 1PL model, as a special case of it), we find the mean of Y and, similarly, its mean in the second equation for the 3PL model. That is, (5.31) expresses the mean of the response variable Y (which can also be viewed as conditional on θ), for any of the three models considered in this chapter, as a nonlinear function of model parameters. Hence, each of these three highly popular and widely used IRT models can actually be seen as a respective (multivariate) nonlinear regression model with a single predictor—the studied trait, ability, or latent continuum, θ—that is not observed. (This fact is the reason why we refer to these models, strictly speaking, as analogs to standard nonlinear regression models, because in the latter, conventionally the predictor is observed; see chapter 12 for the generalization to several unobserved predictors within the framework of multidimensional IRT.) We will keep in mind also this helpful characterization of IRT models in the remainder of the book.

5.8   Chapter conclusion

In this chapter, we further discussed main concepts and relationships in IRT. We were also concerned in more detail with logistic models, in particular the 1PL, 2PL, and 3PL models. We similarly pointed out the analogies that exist between IRT models on the one hand and GLIM, nonlinear factor analysis, logistic regression, and nonlinear regression models on the other hand. These conceptual connections are very helpful, in part because a particularly useful approach to logistic IRT model estimation derives from considering any of these three logistic models as multivariate logistic regression models with a single unobserved predictor—the underlying trait, construct, or ability dimension θ—or with several unobserved predictors as in multidimensional IRT (see chapter 12; compare Cai and Thissen [2015]). This link will also be of relevance in chapter 7, which deals with logistic model fitting and parameter estimation. Based on the present chapter's discussion, we are now ready to move on to empirical applications of IRT, in particular using Stata, as we do in the next chapter.

6 First applications of Stata for item response modeling

As we indicated earlier in the book, there are two main types of item response theory (IRT) models, the logistic and normal ogive models. Because of their mathematical tractability and numerical advantages associated with parameter estimation, in the last 40 years or so, the logistic models have become very popular, at the expense of the normal ogive models. Despite the latter being in a sense the first IRT models, used as early as in the 1940s in the behavioral and educational sciences (for example, Lawley [1943, 1944]), within the last few decades, logistic models have effectively dominated wide-spread empirical IRT utilization. This hegemony became especially evident following the publication of the far-reaching contributions by Birnbaum (1968) in the classic mental test theory book by Lord and Novick (1968). Therefore, in the remainder of the present book, we will exclusively use logistic models.

Starting with this chapter, we will use the Stata command irt, which has been specifically developed for fitting and estimating IRT models. We begin next with a discussion of initial activities needed for this software utilization, such as data reading and obtaining variable descriptives. We then use that command for fitting, estimation, and interpretation of one-parameter logistic (1PL), two-parameter logistic (2PL), and three-parameter logistic (3PL) models. To this end, we make use of a widely circulated dataset that is presented in the file lsat.dta (for example, Rizopoulos [2007]; for convenience, we refer to it as the "LSAT dataset" in the rest of the book). A more detailed discussion on IRT model fitting and parameter estimation is relegated to the next chapter. In the present one, we focus on the use of Stata for logistic model fitting and result interpretation in the frequently encountered setting in empirical research using unidimensional measuring instruments that consist of binary or binary scored items.

6.1 Reading data into Stata and related activities

We commence by reading in (accessing) the above mentioned LSAT dataset with Stata. These data represent the binary responses (0 or 1, interpreted correspondingly as "incorrect" or "correct") obtained from $n = 1000$ subjects on $k = 5$ items. To this end, we use the following Stata command:

```
. use http://www.stata-press.com/data/cirtms/lsat.dta, clear
```

In this command line, the option `clear` after the comma ensures that a dataset accessed in prior conducted activities is no longer available or interferes with our current analyses. (The designation "Path" includes the exact directory location of the file `lsat.dta` on the used computer.) An alternative procedure of reading in a Stata (system) file is via use of the graphical user interface. To achieve this, upon starting Stata, we would point to **File** in the top toolbar. In the drop-down menu, we would choose **Open...** and then select the directory on the computer where that file resides; upon choosing that directory and the relevant data file, Stata will read the latter. Figure 6.1 provides a computer screenshot illustrating this activity using the drop-down menu option.

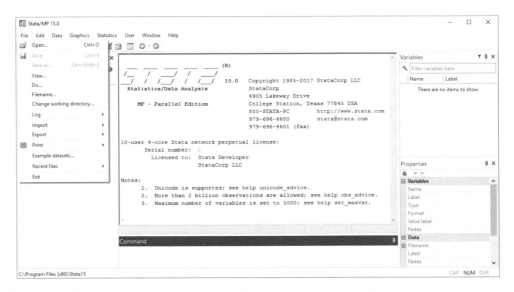

Figure 6.1. Stata computer screenshot of drop-down menu option for accessing a Stata data file

If the LSAT data file were to be available only in an ASCII format (text only format), then one can read it into Stata using the following command (note that the ASCII data file is named `lsat.dat` and has no variable names in its top line; observe also that Stata responds with the statement `1,000 observations read`, as we would expect when correctly reading the data, which is printed by Stata right after the next command):

```
. infile id item1-item5 using http://www.stata-press.com/data/cirtms/lsat.dat
(1,000 observations read)
```

We stress that in the last command, the order of the variables listed immediately after the keyword `infile` must be identical to their order in the raw data file that is being read. (We highly recommend that the information on sample size, number of variables, and their order be available from independent sources before the data file is first accessed with Stata, as is the case in this section.)

Last but not least, to use the graphical user interface option when only an ASCII data file is available, we would point to **File** in the main toolbar and select **Import** and **Unformatted text data**. In the right-most square window, we would press **Browse...** in the top thin window (right end of it) and locate that ASCII file on the computer. We would then type in the second thin window the variable names in the dataset to be accessed, in the order they appear in the latter, check *Replace data in memory* below it, and press **OK** to finally read it in. To now save this dataset in Stata format, for direct access in later analyses, we would point again at **File** in the main toolbar, select **Save as...**, and assign a desired name for the file in a chosen directory that we are prompted to select. (The default extension for Stata data files is `.dta`.) Figure 6.2 provides a computer screenshot illustrating this activity in Stata using the drop-down menu option.

Figure 6.2. Stata computer screenshot of drop-down menu option to access an ASCII file

Once the data file is correctly read in and thus available for further processing by Stata, it is quite useful to list the variable names and related details. We accomplish this with the next command (which can also be shortened to just d):

    . describe

The output produced by Stata is then as follows (as a useful convention indicated earlier, in the rest of this book, we will be presenting the Stata commands in proportionate font and indented, while the output produced by them will be in the same font but slightly smaller and left justified; results generated by a command submitted to the software will be displayed immediately after the former, unless otherwise indicated):

```
Contains data from http://www.stata-press.com/data/cirtms/lsat.dta
  obs:           1,000
  vars:              6                          3 Oct 2016 11:49
  size:         24,000
```

variable name	storage type	display format	value label	variable label
id	float	%9.0g		
item1	float	%9.0g		
item2	float	%9.0g		
item3	float	%9.0g		
item4	float	%9.0g		
item5	float	%9.0g		

```
Sorted by:
```

In this output, Stata informs us that there are 1,000 observations on 6 variables, with both these numbers correctly reflecting the sample size and number of items in the original data file (`lsat.dat` or `lsat.dta`, in the ASCII or Stata format, respectively). The variable or item order, from left to right in that file, is as the one of their listing from top to bottom in the first column of the presented output. In particular, after a subject identifier, denoted `id` as usual, the $k = 5$ dichotomous items follow, which are stored in regular precision. (The remaining columns of this output pertain to data storage and technical details that are not of particular relevance here.)

At this stage, it is important as a matter of routine to make sure that the number of observations and variables in the read data file, which are indicated at the top of the above output, indeed equal the sample size and number of variables in the original dataset, respectively. As mentioned earlier, these two numbers should be known beforehand to the researcher. That sample size and variable number check is accomplished by inspecting the corresponding two numbers in the second and third lines from the top of the output resulting from the Stata command `describe` (or `d`). If the number of observations or variables in the read-in data file does not match the sample size or the number of variables in the initial data file, the reason for this discrepancy needs to be found and corrected before proceeding any further with the next steps outlined below. (That is, an accessed data file with a discrepancy for either of these two numbers should not be processed any further or analyzed until this discrepancy is resolved and the original data file correctly read in with the software.)

A look next at the descriptive statistics of the items can also be informative, in particular prior to commencing the IRT analyses discussed in the next section 6.2. This is achieved with the following command (which can also be shortened to `su`):

```
. summarize item1-item5
```

Variable	Obs	Mean	Std. Dev.	Min	Max
item1	1,000	.924	.2651307	0	1
item2	1,000	.709	.4544508	0	1
item3	1,000	.553	.4974318	0	1
item4	1,000	.763	.4254551	0	1
item5	1,000	.87	.3364717	0	1

By examining the mean of each item, which in this dichotomous item case equals the proportion of "correct" responses, we see from the last presented output that all items are associated with higher than 0.5 probability for that response (denoted "1"). The highest variance is exhibited by item 3, which is because its probability of "correct" response is closest to 0.5 (for example, Raykov and Marcoulides [2011]).

With this initial examination of the dataset of interest, we are now ready to proceed to fitting logistic IRT models and interpreting associated results in the next section. For the specific aims of the present empirical illustration using the LSAT dataset, we assume that each of the five items in it measures a particular aspect of the trait (construct) general mental ability.

6.2   Fitting a two-parameter logistic model

As discussed in chapter 5, the 2PL model represents a general logistic model for empirical settings with no guessing. (We assume initially no guessing on any item in a given set or instrument of interest and will revisit this matter in a following section as well as later in the book.) Therefore, we commence our IRT modeling effort by fitting this model. To this end, we use the following Stata command:

```
. irt 2pl item1-item5
```

This straightforward request yields this output:

```
Fitting fixed-effects model:

Iteration 0:   log likelihood = -2504.5114
Iteration 1:   log likelihood = -2493.5307
Iteration 2:   log likelihood = -2493.4367
Iteration 3:   log likelihood = -2493.4367

Fitting full model:

Iteration 0:   log likelihood = -2478.6867
Iteration 1:   log likelihood = -2467.6539
Iteration 2:   log likelihood = -2466.6565
Iteration 3:   log likelihood = -2466.6536
Iteration 4:   log likelihood = -2466.6536

Two-parameter logistic model              Number of obs    =      1,000
Log likelihood = -2466.6536
```

	Coef.	Std. Err.	z	P>|z|	[95% Conf.	Interval]
item1						
Discrim	.8256703	.2581376	3.20	0.001	.3197299	1.331611
Diff	-3.358777	.8665242	-3.88	0.000	-5.057133	-1.660421
item2						
Discrim	.7227513	.1866698	3.87	0.000	.3568852	1.088618
Diff	-1.370049	.307467	-4.46	0.000	-1.972673	-.7674249
item3						
Discrim	.8907338	.2326049	3.83	0.000	.4348366	1.346631
Diff	-.2796988	.0996259	-2.81	0.005	-.4749621	-.0844356
item4						
Discrim	.6883831	.1851495	3.72	0.000	.3254968	1.05127
Diff	-1.866349	.4343093	-4.30	0.000	-2.71758	-1.015118
item5						
Discrim	.6568946	.2099182	3.13	0.002	.2454624	1.068327
Diff	-3.125751	.8711505	-3.59	0.000	-4.833174	-1.418327

As indicated at the top of this output, after several iterations, the underlying numerical optimization procedure converged to the solution provided above. A main index of relative model fit, which will be used later for comparing the 2PL model with the 1PL model and a 3PL model fit to the same dataset, is the maximized log-likelihood. Its value may be treated, somewhat informally, as a "goodness-of-fit" measure or index that does not explicitly account for model complexity. Hence, this index is best used for model comparison. (For a more detailed discussion, see section 6.5 dealing with nested models and chapter 7 for the data likelihood concept and its maximization.) We note that in the present example, the maximized log-likelihood equals -2466.65 (rounded off).

For each of the five analyzed items, the discrimination and difficulty parameter estimates follow in the subsequently presented panel of the above output. In addition, their associated standard errors, test statistics for being equal to 0 in the population of concern, and pertinent p-values, as well as 95% confidence intervals (CIs), are listed in the remainder of the corresponding rows. Thereby, the information pertaining to the a parameter (item discrimination) precedes that for the b parameter (item difficulty) for each item in this default output layout. (This result presentation layout may arguably be more often of interest in empirical research, but alternative ones are also available; see below.) These findings suggest that under the 2PL model, each item has nonzero discrimination and difficulty parameters in the studied subject population. This interpretation is based on direct inspection of the last 3 columns of above output and, in particular, their CIs, which do not contain the 0 point (suggesting none of these 10 parameters are 0 in the studied population, which as indicated earlier is of relevance with respect to the item discrimination parameters but not the item difficulty parameters).

If one wished a different solution presentation, reordering the lines of the last output is also possible. For instance, if one desired to have the items first "ranked" in terms of their a parameters in ascending order, we request it in the following way:

```
. estat report, byparm sort(a)
```

Two-parameter logistic model   Number of obs   =   1,000
Log likelihood = -2466.6536

	Coef.	Std. Err.	z	P>|z|	[95% Conf. Interval]	
Discrim						
item5	.6568946	.2099182	3.13	0.002	.2454624	1.068327
item4	.6883831	.1851495	3.72	0.000	.3254968	1.05127
item2	.7227513	.1866698	3.87	0.000	.3568852	1.088618
item1	.8256703	.2581376	3.20	0.001	.3197299	1.331611
item3	.8907338	.2326049	3.83	0.000	.4348366	1.346631
Diff						
item5	-3.125751	.8711505	-3.59	0.000	-4.833174	-1.418327
item4	-1.866349	.4343093	-4.30	0.000	-2.71758	-1.015118
item2	-1.370049	.307467	-4.46	0.000	-1.972673	-.7674249
item1	-3.358777	.8665242	-3.88	0.000	-5.057133	-1.660421
item3	-.2796988	.0996259	-2.81	0.005	-.4749621	-.0844356

We note from this output that the only effect of the last-used Stata command is the reordering of the rows in the earlier presented item results section. Hence, none of the results associated with the fit model are changed (because no new model has been fit to the same dataset analyzed). As we can see from the top panel of the last output, in the used sample (dataset), item 5 is the least discriminating one, as judged by the item discrimination parameter estimates. However, given the relatively large standard errors compared with the differences in these estimates for the other items, one cannot suggest from this observation only that item 5 would be the least discriminating item also in the population. In fact, keeping in mind the relatively sizable standard errors, one may as well suggest that the five items have very similar discrimination parameters. This is a potentially rather interesting relationship with respect to the studied population (see also chapter 5). We thus keep in mind this discrimination parameter similarity across items and will pursue it in more detail in the next section.

Alternatively, if one wished instead to have the item difficulty parameter estimates ranked in ascending order, we achieve it this way, observing also that they are all negative here (see lower panel of output below):

```
. estat report, byparm sort(b)
```

Two-parameter logistic model   Number of obs   =   1,000
Log likelihood = -2466.6536

	Coef.	Std. Err.	z	P>|z|	[95% Conf.	Interval]
Discrim						
item1	.8256703	.2581376	3.20	0.001	.3197299	1.331611
item5	.6568946	.2099182	3.13	0.002	.2454624	1.068327
item4	.6883831	.1851495	3.72	0.000	.3254968	1.05127
item2	.7227513	.1866698	3.87	0.000	.3568852	1.088618
item3	.8907338	.2326049	3.83	0.000	.4348366	1.346631
Diff						
item1	-3.358777	.8665242	-3.88	0.000	-5.057133	-1.660421
item5	-3.125751	.8711505	-3.59	0.000	-4.833174	-1.418327
item4	-1.866349	.4343093	-4.30	0.000	-2.71758	-1.015118
item2	-1.370049	.307467	-4.46	0.000	-1.972673	-.7674249
item3	-.2796988	.0996259	-2.81	0.005	-.4749621	-.0844356

However, these 10 parameter estimates (of 5 item discrimination and 5 item difficulty parameters) are not often easy to interpret in purely numeric terms. Thus, we could for instance graph the corresponding item characteristic curves (ICCs) to aid with their interpretation. We achieve it with the following Stata command:

```
. irtgraph icc item1-item5
```

This yields the graph presented in figure 6.3 [with different colors assigned as software default to the different ICCs].

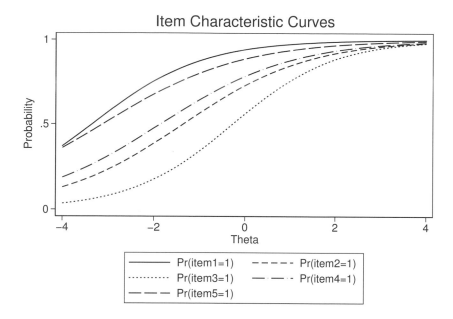

Figure 6.3. Graph of the item characteristic curves for the $k = 5$ analyzed items

We notice from figure 6.3 (see also last presented output) that item 1 is the easiest in the analyzed dataset. To observe this, imagine drawing a horizontal line at the point symbolizing the probability of 0.5 on the vertical axis. Because this line first crosses the ICC of item 1, as one moves from left to right on the horizontal axis, this item is easiest here (see also figure 6.4 below). In the same way, we can also observe that item 3 seems to be the hardest (in this dataset). This is because for item 3, the point on the horizontal axis that corresponds to the intersection of its ICC with that imaginary horizontal line at 0.5 probability is to the right of any such point for the remaining 4 items (see also figure 6.4). In addition, figure 6.3 suggests that the tangents to each ICC at its inflection point (the point of intersection of the ICC with that imaginary horizontal line at 0.5 probability) are possibly fairly close to parallel. Again, to be more confident in such an interpretation, we need additional analyses that we will conduct in section 6.3.

To obtain more precise graphical information about possible item difficulty differences, we can request pointing out the location of the difficulty parameter estimates on the ICC plot. We achieve this with the following command (note that it is the first ICC graphing command with an added subcommand stated after the comma, and see figure 6.4 for the resulting graph):

```
. irtgraph icc item1-item5, blocation
```

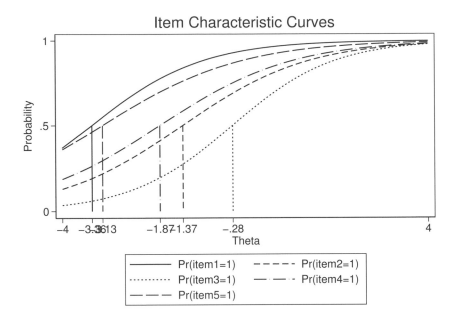

Figure 6.4. Graph of the item characteristic curves for the $k = 5$ analyzed items, with added estimated b parameters (location parameters)

In figure 6.4, we also see that item 3 is the most difficult of all items in the analyzed dataset. To ensure a larger graphical window and better separation of the difficulty parameter estimates along the horizontal axis, we can use this extended Stata command (with resulting graph following it in figure 6.5):

. irtgraph icc, blocation legend(off) xlabel(, alt)

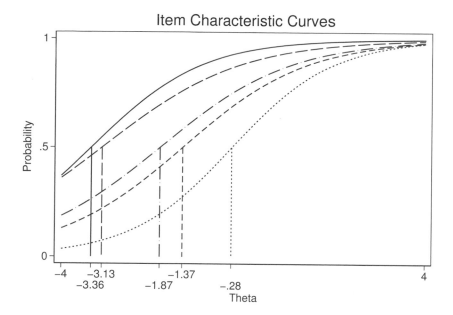

Figure 6.5. Extended graph of the item characteristic curves for the $k = 5$ analyzed items, with alternating indication of the b parameter estimates

To address next the interesting and important question raised above, whether the five ICCs are parallel in their central parts, we need to examine the fit of a more restrictive model that implements this constraint. To this end, we must engage in what is often referred to as nested model testing and model comparison or selection, as we do in the following two sections.

6.3   Testing nested item response theory models and model selection

In applications of IRT in empirical research, frequently a scientist is interested in examining two or more models that are fit to the same dataset (items). For instance, in section 6.2, we fit a 2PL model, but when we inspect its solution, it appeared sensible to evaluate also a more restrictive model that stipulates the identity of all item discrimination parameters. This restriction is an integral part of the 1PL model (Rasch model), as mentioned earlier in the book (see chapter 5). When imposed in the 2PL model, this item parameter constraint renders from the latter that more restrictive model, the 1PL model. As will be recalled, a pair or sequence of more than two models is called nested if i) they are all fit to the same dataset (observed variables or items) and ii) except one model, any of them is obtained from another model in the sequence by placing constraints on some of its parameters. Nested models are very widely used in applied statistics, and in particular to test aspects of substantive theories or hypotheses of interest that are represented by the nesting restrictions (parameter constraints).

In the currently considered IRT setting, the 1PL model is nested in the 2PL model as mentioned earlier. Indeed, the former model is obtained from the latter model by introducing the restriction

$$a_1 = a_2 = \cdots = a_k \tag{6.1}$$

for the item discrimination parameters, with k denoting the number of items in a set or measuring instrument of interest ($k > 1$).

To examine empirically the plausibility of this restriction of item-invariant discrimination power, which characterizes the 1PL model (Rasch model), we can make use of the popular likelihood-ratio test (LRT) (for example, Johnson and Wichern [2007]). As its name suggests, the LRT compares the maximized log-likelihoods of two nested models for a given set of observed variables (items) to which they are fit. As we indicated in the preceding section, the maximized log-likelihood for a model fit using the method of maximum likelihood can be viewed informally as a measure of its fit to the data, which does not account for model complexity. In particular, one can further maximize the log-likelihood by including additional parameters. It is unclear in general, however, whether this improvement in fit is "real" (significant) in the population, that is, if it renders the more complex model as a better means of data description and explanation. Hence, whether the nesting restriction is plausible in the studied population, and thus there is no "real" (significant) loss of fit when moving from the more complex to the more restrictive model, is actually tested by comparing the maximized log-likelihoods across the two models. This is accomplished with a corresponding application of the LRT.

The LRT statistic for testing a pair of nested models is defined as follows:

$$\Delta l = -2(l_0^* - l_1^*) \tag{6.2}$$

In (6.2), l_0^* is the maximized log-likelihood of the model after the restriction is introduced (that is, of the nested or restricted model, sometimes also called the "null" model). Further, l_1^* is the maximized log-likelihood of the model before that restriction is imposed (sometimes called the "alternative" model). The test statistic (6.2) is distributed with large samples, under the tested null hypothesis H_0: "The introduced nesting constraint(s) is correct in the population", like a chi-squared random variable with degrees of freedom being the difference $q_1 - q_0$ in the number of (free) parameters involved in both models (for example, Johnson and Wichern [2007]). (The more general or less restrictive model is frequently also called the full model, and the nested or restricted model is referred to as the reduced model.) We will use the LRT on a number of occasions in the remainder of the book to examine various nested models and test parameter restrictions.

While the LRT is a statistical test that is applicable with nested models, sometimes a researcher may be interested in models that are not necessarily nested while still being fit to the same dataset (observed variables). In that case, given the above mentioned feature of the maximized log-likelihood improving by adding parameters, it would be desirable to introduce some penalty for model complexity, in particular for the number of model parameters. In this way, a more veridical comparison of the models can be achieved. This comparison accounts for the above mentioned fact that increased maximized log-likelihood may be obtained just by adding parameters to an initial model without really improving its data fit in an important and meaningful way.

Two widely used model comparison indices that implement this idea of penalizing for increased model complexity are the Akaike information criterion (AIC) and the Bayesian information criterion (BIC) (for example Raftery [1995]). They are also often referred to as information criteria (indices). The AIC is defined as

$$\text{AIC} = -2l^* + 2q \tag{6.3}$$

and the BIC as

$$\text{BIC} = -2l^* + q\ln(n) \tag{6.4}$$

In (6.3) and (6.4), q is the number of (free) parameters in a model of interest, the maximized log-likelihood in it is denoted l^*, and n is sample size. We note that while the AIC introduces the penalty $2q$ for model complexity, the BIC uses for it the product $q\ln(n)$. Other than this penalty, both information criteria are based on (-2) times the maximized log-likelihood for the model of concern in their definition formulas.

The AIC and BIC are used mostly for model comparison and cannot be meaningfully interpreted when only a single model is fit to a given dataset. (It may well be useful, however, to compute these indices for an initial model and "keep them on file", for the case of considering later another model fit to the same data, as we do next.) In particular, among a set of rival models fit to the same observed variables and dataset (with the models possibly not being nested), the one associated with the smallest AIC and BIC is considered the preferred model (for example, Raftery [1995]).

For the 2PL model fit in section 6.2, we can readily obtain its AIC and BIC using direct computations with the command **generate**, which can be shortened to **gen**. To this end, first we need to find out the number of parameters in this model. Because in it each of the $k = 5$ items has a discrimination and difficulty parameter, there is a total of $q = 5 + 5 = 10$ parameters. (We note that in simple terms, this is the number of quantities in the pertinent output, which are associated with standard errors; see section 6.2.)

Recalling that sample size is $n = 1000$ here, we can now calculate the AIC and BIC indexes for the 2PL model by using the following Stata commands (with the third command line, we list only the pair of evaluated indices):

```
. generate AIC = (-2)*(-2466.6536) + 2*10
. generate BIC = (-2)*(-2466.6536) + 10*log(1000)
```

```
. list AIC BIC in 1/1
```

These three commands yield the following output:

	AIC	BIC
1.	4953.307	5002.385

Because the AIC and BIC indices are so widely used for model comparison—a topic that we will revisit in the next chapter—their computation has been already implemented in Stata. Indeed, it is invoked in this software with what is called a postestimation command, which is a command applied on the results of a prior estimation process (model fitting). This command is `estat ic`, and we can use it here right after fitting the 2PL model under consideration (output presented immediately after command):

```
. estat ic
```
Akaike´s information criterion and Bayesian information criterion

Model	Obs	ll(null)	ll(model)	df	AIC	BIC
.	1,000	.	-2466.654	10	4953.307	5002.385

Note: N=Obs used in calculating BIC; see [R] BIC note.

As expected, the AIC and BIC results are identical to those obtained earlier with their computational commands based on (6.3) and (6.4). (The column titled `df` in the output is actually meant to represent the number q of model parameters.) We keep this information handy for our discussion in the next two sections. In particular, we will compare these two indices with the same information criteria of two other models fit to the five items of the LSAT dataset that is used throughout this chapter for illustration purposes.

6.4 Fitting a one-parameter logistic model and comparison with the two-parameter logistic model

As we found in section 6.2, when fitting the 2PL model, the five analyzed items seemed to function very similarly in terms of discriminating between persons with ability levels around their ICC inflection point projections on the horizontal axis or θ-scale (that is, around their difficulty parameters). This suggested that perhaps the 1PL model, that is, the Rasch model, may fit the data "almost" as well as the 2PL model (see also section 6.3). In the present section, we pursue this suggestion concerning these two popular IRT models when applied on the used LSAT dataset.

To fit the 1PL model with Stata, we use the following command (while still having the same data file `lsat.dta` active, that is, accessed by the software):

```
. irt 1pl item1-item5
Fitting fixed-effects model:
Iteration 0:   log likelihood = -2504.5114
Iteration 1:   log likelihood = -2493.5307
Iteration 2:   log likelihood = -2493.4367
Iteration 3:   log likelihood = -2493.4367
Fitting full model:
Iteration 0:   log likelihood = -2472.1374
Iteration 1:   log likelihood = -2467.0094
Iteration 2:   log likelihood = -2466.9376
Iteration 3:   log likelihood = -2466.9376
```

| One-parameter logistic model | | | | | Number of obs | = | 1,000 |
| Log likelihood = -2466.9376 | | | | | | | |

	Coef.	Std. Err.	z	P>|z|	[95% Conf.	Interval]
Discrim	.7551283	.0694206	10.88	0.000	.6190666	.8911901
item1						
Diff	-3.615293	.3265991	-11.07	0.000	-4.255416	-2.975171
item2						
Diff	-1.322434	.1421673	-9.30	0.000	-1.601077	-1.043792
item3						
Diff	-.3176353	.0976766	-3.25	0.001	-.5090779	-.1261928
item4						
Diff	-1.730106	.1691149	-10.23	0.000	-2.061565	-1.398647
item5						
Diff	-2.780193	.251015	-11.08	0.000	-3.272174	-2.288213

We notice that the maximized log-likelihood for the Rasch model, being -2466.9376, is only marginally worse than this quantity for the 2PL model fit in section 6.2. That the maximized log-likelihood of the latter model is higher should not come as a surprise at all. The reason is that the 1PL model has four fewer parameters, because of its assumption that all five item discrimination parameters are the same, and hence is more restrictive. Thus, its maximized log-likelihood cannot be higher than that of the 2PL model.

However, the actual question here is whether this difference in the log-likelihoods is "real" or "significant". More specifically, we ask whether this difference is explainable by chance fluctuations only, which are typically at work when drawing a sample from a studied (subject) population. To answer this query, we use the LRT, which as mentioned above is widely used in applied statistics as a means of comparing two nested models, such as the 1PL and 2PL models here (see preceding discussion in this chapter). Hence, the LRT provides a test statistic of the restrictions in (6.1) that are the only aspect in which the 1PL model and 2PL model differ. As indicated earlier, the parameter constraints in (6.1) in fact nest the more restrictive model into the more general model, that is, the Rasch model into the 2PL model.

102   *Chapter 6   First applications of Stata for item response modeling*

The LRT is readily invoked with Stata in the present IRT context. To this end, we need three initial commands, which are stated next, to prepare for this test that is conducted with the last command after them:

```
. quietly irt 2pl item1-item5
. estimate store m2pl
. quietly irt 1pl item1-item5
. lrtest m2pl
```

In this sequence of four commands, first we request refitting the 2PL model but suppress printing of the results to the computer screen, because we have already inspected them in section 6.2, where we first fit that model. (Thus, we prefix the first `irt` command with `quietly`.) Then, with the second command line, we store the results associated with the fit 2PL model in what may be called an object named m2pl. We next fit with the third command line the 1PL model, suppressing again the associated output because we are already familiar with it from the preceding discussion in this section. Finally, to conduct the LRT, which consists here of comparing the above mentioned fit statistics of the nested model (1PL model) with that of the full model (2PL model) [see (6.2), we use the command `lrtest`, which is succeeded by the object created for the more general model. (The default treatment by Stata is of the m2pl object as pertaining to the less restrictive model fit before the one nested in it.)

The discussed four-command sequence above produces this output:

```
Likelihood-ratio test                         LR chi2(4)  =       0.57
(Assumption: . nested in m2pl)                 Prob > chi2 =     0.9666
```

According to these results, because the p-value associated with the LRT is far from being significant (using any reasonable significance level), we conclude that the constraint of equal item discrimination parameters for the five analyzed items that was imposed in the 1PL model is plausible in the studied population. (The symbol . in the second line on the left in this output refers to the last fit model.) That is, there is no sufficient evidence in the analyzed data to warrant rejection of the tested null hypothesis of equality of the 5 item discrimination parameters [the next set of equations is identical to (6.1) for the $k = 5$ items of relevance here]:

$$H_0: a_1 = a_2 = \cdots = a_5$$

This result suggests that if one were to choose between the 2PL model and the 1PL model (Rasch model), one could reason in the following way. Given the last test findings, one would prefer the Rasch model on grounds of i) essentially the same fit to the data (specifically, insignificantly worse fit to the analyzed data as reflected in the maximized log-likelihood); and ii) higher parsimony, that is, fewer parameters.

In addition to the LRT, as mentioned in section 6.3, it would be also useful to examine the information criteria associated with both models. We indicated earlier in the chapter that we can obtain them for either model by using the Stata command `estat ic`. To this end, and for completeness of the current discussion in this section, we refit both models, suppress their output, and request their information criteria:

```
. quietly irt 2pl item1-item5
. estat ic
Akaike's information criterion and Bayesian information criterion
```

Model	Obs	ll(null)	ll(model)	df	AIC	BIC
.	1,000	.	-2466.654	10	4953.307	5002.385

Note: N=Obs used in calculating BIC; see [R] BIC note.

```
. quietly irt 1pl item1-item5
. estat ic
Akaike's information criterion and Bayesian information criterion
```

Model	Obs	ll(null)	ll(model)	df	AIC	BIC
.	1,000	.	-2466.938	6	4945.875	4975.322

Note: N=Obs used in calculating BIC; see [R] BIC note.

As we see from these results, and specifically from the last two columns of the outputs, the 1PL model has considerably lower BIC and AIC. Hence, the 1PL model can also be preferred to the 2PL model based on these information criteria (for example, Raftery [1995]).

Now that we have settled on the 1PL model, in the remainder of this section, we discuss its results further. First, we can request a presentation of its associated output, including a ranking of the difficulty parameters (because the item discrimination parameters are all the same in this model). We achieve it using the following command:

```
. estat report, byparm sort(b)
One-parameter logistic model                 Number of obs    =       1,000
Log likelihood = -2466.9376
```

	Coef.	Std. Err.	z	P>|z|	[95% Conf. Interval]
Discrim	.7551283	.0694206	10.88	0.000	.6190666 .8911901
Diff					
item1	-3.615293	.3265991	-11.07	0.000	-4.255416 -2.975171
item5	-2.780193	.251015	-11.08	0.000	-3.272174 -2.288213
item4	-1.730106	.1691149	-10.23	0.000	-2.061565 -1.398647
item2	-1.322434	.1421673	-9.30	0.000	-1.601077 -1.043792
item3	-.3176353	.0976766	-3.25	0.001	-.5090779 -.1261928

The last listing of the items, from top to bottom, presents the ranking of their difficulty parameter estimates, along with their 95% CIs. This listing indicates for the analyzed dataset that i) item 1 is easiest because its b parameter estimate is the smallest, and ii) item 3 is the most difficult because its b parameter estimate is the largest. To see the item differences even more clearly in this 1PL model, we take a look at the (estimated) ICCs within it, requesting printing of their difficulty parameter estimates. All this information is presented in figure 6.6 (see also section 6.2 for the pertinent command, and observe the subcommand added next after the comma):

```
. irtgraph icc item1-item5, blocation
```

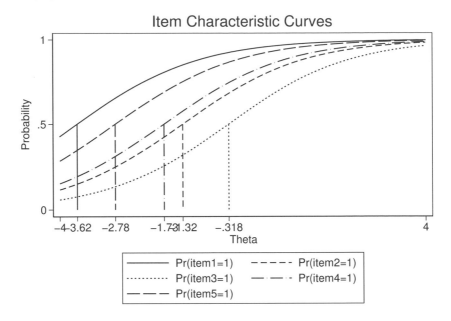

Figure 6.6. Graph of the item characteristic curves for the $k = 5$ analyzed items using the one-parameter logistic model (Rasch model) with added location indication for the b parameters estimates

Figure 6.6 lets us also easily notice the parallelism feature of the ICCs. Indeed, their imaginary tangents at the respective inflection points are all parallel. In addition, the above output and figure allow us to observe more clearly, in tandem with the last mentioned finding, the differences between the items in terms of their difficulty parameters relative to the 2PL model. This is because the Rasch model, which we preferred to the 2PL model here, is more parsimonious. Therefore, it is associated with smaller standard errors and narrower CIs. This fact implies higher estimation precision and thus clearer parameter comparison. In particular, we see that item 3 has an estimated difficulty parameter notably above all other items (in the analyzed data). Hence, this item requires the highest level of the presumably evaluated general mental ability with this 5-item instrument for achieving a probability of 0.5 for its "correct" answer. That is, item 3 appears to be the most difficult item here. In addition, figure 6.6 graphically illustrates the highly useful feature of IRT of positioning (estimating) the items on the same dimension that represents the studied trait, construct, or ability, that is, on the horizontal axis in this figure (compare van der Linden [2016b]). This feature allows, as mentioned earlier, locating both items (item difficulty parameters) and examined persons on the same continuum, the one denoted typically θ and frequently referred to as "θ-scale".

We stress that the discussed item difficulty differences in the fit Rasch model are only of descriptive nature. The reason is that they were observed merely in the respective sample estimates from the analyzed dataset. In addition, we notice that the standard errors for these parameters are of similar magnitude as the *b*-parameter estimate differences across the five items. This makes their earlier difficulty parameter comparisons even more speculative. To shed more light on these possible item differences, we can use the difficulty parameters' CIs and in particular examine if they are overlapping (see also footnote 1 below). Comparing these CIs for items 1 and 4 that are not overlapping, it is suggested that the difficulty of item 1 may be lower than that of item 4 in the studied population. This is because the CI for the difficulty parameter of item 1 is entirely to the left of the CI for that parameter of item 4. By the same token, item 1 is also seen as easier than item 3, because the CI for the difficulty parameter of the latter item is entirely to the right of the corresponding CI for the former item. Similarly, we find that item 2 is easier than item 3 but more difficult than each of items 1 and 5. However, the potential difference in difficulty between item 2 and item 4 is less clear if at all the case in the population, like that difference in difficulty between items 1 and 5. This is due to the considerable overlap of the pertinent CIs, as seen from the last two columns of the above output with parameter estimates in the 1PL model. Last but not least, item 3 is suggested with this CI-based comparison to be the most difficult in the entire instrument (in the analyzed dataset). Indeed, its difficulty parameter's CI is to the right of the respective CIs of all other items. We point out, however, that these CI comparisons contain a somewhat informal element and cannot be considered equivalent to a statistical test or a series of such for addressing the same comparison questions about population item difficulty parameters. Such tests can be conducted using the LRT in analogy to the earlier developments in this section.[1] (See also, for example, Raykov and Marcoulides [2013] for a possible one-tailed testing approach, if it is decided for, before looking at the data.)

---

1. If interested in examining the difference in two item parameters, such as the difficulty parameters of two given items, you may use a CI for it (or a closely related test of the null hypothesis of their identity versus a two-tailed alternative). This interval is readily obtained as follows. First, using the command `irt, coeflegend` immediately after fitting the model of concern, find out the internal reference by the software to the parameters in question. For instance, if you are interested in interval estimation of the difference in the difficulty parameters of items 2 and 3 for the used LSAT example, the last stated command determines that they are referred to internally as `_b[item2:_cons]` and `_b[item3:_cons]`, respectively, while the common discrimination parameter is referred to as `_b[item2:Theta]`. A 95% CI of the difference $b_2 - b_3$ in these difficulty parameters is then obtained with the command `nlcom _b[item2:_cons]/(-_b[item2:Theta])-_b[item3:_cons]/(-_b[item3:Theta])` and is found to be $(-1.30, -0.71)$ (see chapter 4 for the intercept-slope parameterization involved in fitting this 1PL model; see also the *Stata Item Response Theory Reference Manual* [2017a]). This interval can be interpreted as suggesting that an interval consisting of plausible values for this difference in the studied population, with high confidence, stretches from -1.30 through -0.71 and hence consists entirely of negative values. This is consistent with item 2 being less difficult than item 3 in the population. Because this interval does not include 0, if one was instead interested in testing their difference to begin with, one would reject the null hypothesis of their identity at the 0.05 significance level (for example, Raykov and Marcoulides [2008]). (Note that the outlined method of parameter difference interval estimation is applicable along the same lines also with other IRT models used in later chapters of the book.) The LRT mentioned in the main text would instead fit the nested model with these parameters constrained for equality and compare its maximized log-likelihood, as discussed earlier in the chapter, with that of the 1PL model without this restriction.

With this discussion, we are now ready to compare either of the two IRT models fit so far with an alternative model, which allows for guessing on the items under consideration.

6.5 Fitting a three-parameter logistic model and comparison with more parsimonious models

In some educational and in particular ability assessment contexts, a 3PL model may be appropriate because of possible "guessing" by some studied persons on one or more items (see also chapter 5). Such individuals may tend to possess low ability levels and elect to "guess" the correct answers on them. To respond to this challenge, as discussed previously in the chapter, the 3PL model includes a third parameter for any item on which "guessing" is suspected.

Although one could obviously argue that it is subjects rather than items that "guess", since the 1960s, the 3PL model has been widely applied in settings where chance-based decisions on part of studied persons may lead to them selecting the correct answer on an item of concern even in the absence of a sufficient level of the ability (presumably) needed for finding that answer. Unfortunately, and despite its use since the 1960s, the 3PL model is frequently associated with numerical difficulties involving in particular lack of convergence. In such cases, one possible avenue to pursue to resolve this challenge is to adopt a Bayesian statistics approach (for example, Cai and Thissen [2015]). That approach is particularly attractive when it is assumed that items with suspected guessing have their own distinct pseudoguessing parameters. Alternatively, a 3PL model with a common pseudoguessing parameter can be fit using essentially the same type of estimation approach as with the more parsimonious logistic models used earlier in the chapter. In this section, we discuss the latter 3PL model and illustrate it with empirical data. We return to the more general 3PL model in a later chapter of the book (chapter 11), where we use a hybrid IRT modeling approach to relax this assumption of pseudoguessing parameter identity across items.

To fit to the LSAT dataset a 3PL model with a pseudoguessing parameter common to all items, we can use the following Stata command:

```
. irt 3pl item1-item5

Fitting fixed-effects model:

Iteration 0:    log likelihood = -2799.0926
Iteration 1:    log likelihood = -2502.1504   (not concave)
Iteration 2:    log likelihood =  -2495.174
Iteration 3:    log likelihood = -2493.4369   (not concave)
Iteration 4:    log likelihood = -2493.4367   (backed up)

Fitting full model:

Iteration 0:    log likelihood = -2477.8707
Iteration 1:    log likelihood = -2475.9738   (not concave)
Iteration 2:    log likelihood = -2475.4904   (not concave)
Iteration 3:    log likelihood = -2468.3078   (not concave)
Iteration 4:    log likelihood = -2467.5869   (not concave)
```

```
Iteration 5:   log likelihood =   -2467.22 (not concave)
Iteration 6:   log likelihood = -2467.0534 (not concave)
Iteration 7:   log likelihood = -2466.9112 (not concave)
Iteration 8:   log likelihood = -2466.8452 (not concave)
Iteration 9:   log likelihood = -2466.8087
Iteration 10:  log likelihood = -2466.7737
Iteration 11:  log likelihood = -2466.7314
Iteration 12:  log likelihood = -2466.7281
Iteration 13:  log likelihood = -2466.6942
Iteration 14:  log likelihood = -2466.6855
Iteration 15:  log likelihood = -2466.6755
Iteration 16:  log likelihood = -2466.6681
Iteration 17:  log likelihood = -2466.6649
Iteration 18:  log likelihood = -2466.6606
Iteration 19:  log likelihood = -2466.6564 (not concave)
Iteration 20:  log likelihood = -2466.6563
Iteration 21:  log likelihood = -2466.6557
Iteration 22:  log likelihood =  -2466.655
Iteration 23:  log likelihood = -2466.6543 (not concave)
Iteration 24:  log likelihood = -2466.6543
Iteration 25:  log likelihood = -2466.6541
Iteration 26:  log likelihood = -2466.6538
Iteration 27:  log likelihood = -2466.6538
Iteration 28:  log likelihood = -2466.6537
Iteration 29:  log likelihood = -2466.6536
Iteration 30:  log likelihood = -2466.6536
```

```
Three-parameter logistic model              Number of obs   =      1,000
Log likelihood = -2466.6536
```

	Coef.	Std. Err.	z	P>|z|	[95% Conf.	Interval]
item1						
Discrim	.8256751	.2581351	3.20	0.001	.3197396	1.331611
Diff	-3.358698	.8666539	-3.88	0.000	-5.057308	-1.660088
item2						
Discrim	.7227716	.1867226	3.87	0.000	.356802	1.088741
Diff	-1.36992	.3087103	-4.44	0.000	-1.974981	-.7648587
item3						
Discrim	.890782	.232855	3.83	0.000	.4343946	1.347169
Diff	-.2795792	.1029265	-2.72	0.007	-.4813115	-.077847
item4						
Discrim	.6883953	.1851667	3.72	0.000	.3254753	1.051315
Diff	-1.866228	.4350721	-4.29	0.000	-2.718953	-1.013502
item5						
Discrim	.6569029	.2099225	3.13	0.002	.2454624	1.068343
Diff	-3.125633	.8714679	-3.59	0.000	-4.833679	-1.417587
Guess	.0000223	.0098813	0.00	0.998	-.0193447	.0193893

We see that fitting this 3PL model involved considerably more iterations as well as some initial numerical difficulties. The latter are indicated by the keywords "not concave" and "backed up" and are associated with a less than uneventful early iteration development. More importantly, however, after those initial issues the numerical procedure stabilized and converged to the solution presented. We also notice that the fit model assumes a single pseudoguessing parameter associated with all items. The estimate of that parameter is presented in the last line of the above output. This default setting in Stata reflects the previously mentioned general difficulty of estimating 3PL models. We similarly notice that this item-invariant parameter is not significant. This is because its *p*-value is much higher than a reasonable (preset) significance level, and its 95% CI covers the 0 point (see, for example, Raykov and Marcoulides [2008], on what may be seen as "duality" between using for some hypotheses the conventional testing procedure or associated CIs). This test of the null hypothesis $H_0: g = 0$ for the items, that is, vanishing common pseudoguessing population parameter, is an example of conducting what is usually referred to as a Wald test of a single model parameter. The test is based on a direct numerical comparison (division) of the parameter estimate with its standard error. This renders a test statistic that is asymptotically distributed like a standard normal variate under the null hypothesis of that parameter being 0 in the population (for example, Baker and Kim [2004]).

In addition to this test of the common pseudoguessing parameter, for the aim of model choice, we can also compare the information criteria of this 3PL model with those of the 1PL and 2PL models fit earlier in the chapter. As illustrated before, we obtain them using the following command (immediately after the 3PL model in question is fit to data):

```
. estat ic
Akaike´s information criterion and Bayesian information criterion
```

Model	Obs	ll(null)	ll(model)	df	AIC	BIC
.	1,000	.	-2466.654	11	4955.307	5009.293

Note: N=Obs used in calculating BIC; see [R] BIC note.

We observe that both the AIC and BIC are notably higher in the considered 3PL model relative to the Rasch and 2PL models fit previously. This leads to the suggestion that either of the latter two models would be selected when compared with that 3PL model. Based on these comparisons, as well as our earlier discussion on the 1PL and 2PL model for the used LSAT dataset, we can conclude that the Rasch model is preferable to the 2PL model as well as to the 3PL model with an item-invariant pseudoguessing parameter.[2]

We will revisit the 3PL model later in the book. Specifically, in chapter 11, we will relax its item-invariance assumption for the pseudoguessing parameter used in the present section and will use a hybrid IRT modeling approach to fit the resulting more general 3PL model.

6.6 Estimation of individual subject trait, construct, or ability levels

A particular feature of IRT, which we mentioned earlier (for example, chapter 1), is that it provides also the opportunity to evaluate individual person levels of the studied trait or ability, often denoted $\theta_1, \ldots, \theta_n$ (with n symbolizing sample size). These are the particular unknown positions on the latent continuum of actual interest, at which the examined individuals are located.

In this property of offering the opportunity to evaluate (estimate, predict) also these individual positions, IRT may be seen at first as somewhat distinct from other applied statistical frameworks used for estimation of model parameters, such as, say, the intercept and partial regression coefficients in a regression model. As is well known, however, standard regression analysis can be used to predict for a given subject his or her unknown dependent variable value using their observations on the explanatory variables and a plausible model for their relationships (recall that this prediction is also associated with a standard error; see, for example, Agresti and Finlay [2009]). Similarly,

---

2. That the 1PL model happens to be the preferred model for the used LSAT dataset in this chapter should not be interpreted as implying (nor is it meant to imply) that the end goal of measurement has to be finding a set of items—from, say, an initial larger group of items under consideration—that comply with the Rasch model (compare Raykov and Calantone [2014]). The bottom line of behavioral, educational, and social measurement cannot be anything other than validity, especially construct validity, as the most encompassing form of validity (compare Raykov and Marcoulides [2011]). In particular, measurement in these sciences and well beyond cannot be about engaging in activities leading to a particular parameter constraint—like that in (5.16) or (5.22) in chapter 5 stating item-invariant discrimination parameters—being plausible for a given dataset, at times possibly at the expense of dropping initially considered items and thus potentially compromising validity, thus possibly entailing also construct underrepresentation (for example, Messick [1995]). Also, the following simple argument suggests that setting as a goal of behavioral, educational, or social measurement the satisfaction of the Rasch model is difficult to accept logically or from a philosophy of science viewpoint: whether (5.16) is found to be fulfilled (that is, the pertinent null hypothesis not rejected) when tested in an empirical study on a population under investigation, and thus the characteristic Rasch model feature to be satisfied (plausible), is obviously also a function of sample size when this critical requirement of the 1PL model is tested against data or evaluated using statistical tests.

we can estimate the unknown scores of individual persons in a factor analysis model when plausible for a given dataset, that is, obtain factor score estimates for each studied person after fitting the model. (Recall also that these estimates are similarly associated with standard errors; see, for example, Johnson and Wichern [2007].) Incidentally, one often used method for accomplishing this aim, referred to as the regression method for factor score estimation, is based on main ideas underlying traditional regression analysis as well (for example, Raykov and Marcoulides [2008]).

Estimation of the individual person scores on the latent unobserved trait or ability dimension in IRT, which are also considered parameters within this applied statistical framework, can and often will be an essential part of its application in an empirical setting. In fact, one may see this individual trait-level estimation as analogous to the above regression model-based prediction. Indeed, one could surmise this analogy by looking again at figure 1.2 in chapter 1. As we alluded to there, we would like to use in item response modeling the data on the observed items to make inferences with respect also to the unknown individual person positions or locations (scores) on the underlying unobserved trait dimension, construct, or latent continuum, that is, on the θ-scale.

This use of the observed data on the items to furnish estimates (predictions) of the individual scores on the trait or ability dimension (or dimensions) of special concern in IRT is the essence of the process of estimation (prediction) of these scores in empirical research. While it will be addressed further in the next chapter, in the rest of this section, we will illustrate this process using Stata. With this estimation procedure, once we have selected a model that is plausible for a given dataset, we can readily obtain with Stata those individual estimates (predictions or assigned values), along with a standard error for each one of them. We will denote these estimates generically $\widehat{\theta}$ throughout the rest of the book.

To obtain the individual trait or ability level estimates in a given dataset, symbolized by $\widehat{\theta}_1$ through $\widehat{\theta}_n$, we can readily use Stata. Thereby, the following command leads to these estimates being stored in the newly constructed variable `Theta`, as well as their associated standard errors in the variable `ThetaSE`:

```
. predict Theta, latent se(ThetaSE)
(option ebmeans assumed)
(using 7 quadrature points)
```

For the empirical example with the LSAT data used earlier in this chapter, as was found before, we preferred the Rasch model i) to the 2PL model on grounds of their fits to the data being very similar and the former being more parsimonious than the latter; and ii) to the 3PL model with a common pseudoguessing parameter in the last section (see chapter 11 for a more general treatment of this model). To ensure we estimate (predict) the individual trait levels associated with this Rasch model, namely, $\theta_1, \ldots, \theta_n$, next we `quietly` fit it first and then request these estimates with the above command, which is stated subsequently:

```
. quietly irt 1pl item1-item5

. predict Theta, latent se(ThetaSE)
(option ebmeans assumed)
(using 7 quadrature points)
```

As seen from its response, Stata used (by default) what is referred to as empirical Bayes means for the estimation or prediction of the individual trait levels, $\theta_1, \ldots, \theta_n$, using 7 quadrature points, a numerical algorithm implementation detail (see *Stata Item Response Theory Reference Manual* [2017a]). Upon inspecting now the analyzed data file—here the Stata active file `lsat.dta`—we can make an interesting observation. Specifically, as a final column, the standard errors of these individual estimates are included in the file, and as a penultimate column, the actual individual trait or ability estimates were added internally by the software. Indeed, next using the `describe` command, which as mentioned earlier provides us among other things with the names of the variables in the opened file, we also notice the final Stata statement that the active dataset has been also changed accordingly:

```
. describe
Contains data from http://www.stata-press.com/data/cirtms/lsat.dta
  obs:          1,000
  vars:             8                          3 Oct 2016 11:49
  size:        32,000
```

variable name	storage type	display format	value label	variable label
id	float	%9.0g		
item1	float	%9.0g		
item2	float	%9.0g		
item3	float	%9.0g		
item4	float	%9.0g		
item5	float	%9.0g		
ThetaSE	float	%9.0g		S.E. of empirical Bayes means for Theta
Theta	float	%9.0g		empirical Bayes means for Theta

```
Sorted by:
     Note: Dataset has changed since last saved.
```

Next, if we wanted to take a look at the first, say, 20 subjects' trait or ability level estimates along with their standard errors (or alternatively for other persons from the original file), we could list them with this command:

```
. list in 1/20
```

	id	item1	item2	item3	item4	item5	ThetaSE	Theta
1.	1	0	0	0	0	0	.7973221	-1.910115
2.	2	0	0	0	0	0	.7973221	-1.910115
3.	3	0	0	0	0	0	.7973221	-1.910115
4.	4	0	0	0	0	1	.8003169	-1.428806
5.	5	0	0	0	0	1	.8003169	-1.428806
6.	6	0	0	0	0	1	.8003169	-1.428806
7.	7	0	0	0	0	1	.8003169	-1.428806
8.	8	0	0	0	0	1	.8003169	-1.428806
9.	9	0	0	0	0	1	.8003169	-1.428806
10.	10	0	0	0	1	0	.8003169	-1.428806
11.	11	0	0	0	1	0	.8003169	-1.428806
12.	12	0	0	0	1	1	.8087799	-.9405584
13.	13	0	0	0	1	1	.8087799	-.9405584
14.	14	0	0	0	1	1	.8087799	-.9405584
15.	15	0	0	0	1	1	.8087799	-.9405584
16.	16	0	0	0	1	1	.8087799	-.9405584
17.	17	0	0	0	1	1	.8087799	-.9405584
18.	18	0	0	0	1	1	.8087799	-.9405584
19.	19	0	0	0	1	1	.8087799	-.9405584
20.	20	0	0	0	1	1	.8087799	-.9405584

As we can see from the output, subjects (among the 20 listed) who have the same number of "correct" responses, or 1s, on the 5 items that were administered to them have been assigned the same individual trait or ability level estimate and are associated with the same standard error. This is related to a characteristic property of the Rasch model that the number of correct scores (that is, the sum of the item responses here) is a sufficient statistic in it for the underlying individual subject trait level (see, for example, von Davier [2016]; see also Reckase [2009]).

The discussed, extended LSAT data file includes the individual ability level estimates and standard errors in numerical form and in this way makes them easily available for subsequent analyses if needed. In addition, at times, it may also be of interest—especially for individual person comparison aims—to graph these estimates along with appropriate functions of their standard errors, say, presented in the vertical direction. To this end, one can proceed as follows. First, we can add and subtract 1.96 times the standard error to each latent score estimate to obtain its individual 95% CI, based on the large-sample normality of these estimates (see chapter 7). We achieve this in the following way:

```
. generate lower = Theta - 1.96*ThetaSE
. generate upper = Theta + 1.96*ThetaSE
```

To graph next the subject trait level estimates along with their CIs positioned vertically, we use the following command (see figure 6.7 for the resulting plot):

```
. twoway rcap upper lower id || scatter Theta id
```

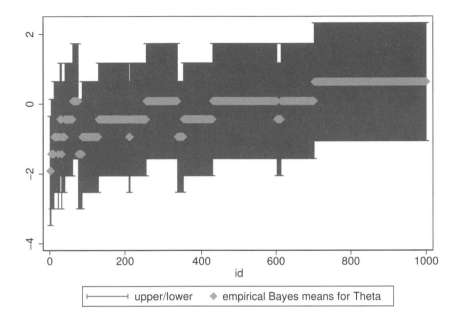

Figure 6.7. Graph of individual latent score estimates (predictions) with their 95% confidence intervals (in vertical direction)

This graph is clearly "overpopulated" because it presents the assigned subjects' latent score estimates (predictions) and associated CIs for the total sample of $n = 1000$ examined persons. However, we can still see from it that some of the persons with lower identifiers (IDs) have their CIs notably below those of subjects with higher identifiers. Given that the observations (cases) have been actually sorted in ascending order in terms of number correct in the original dataset to begin with, if of interest one could select a few subjects from the top and from the bottom, say, to examine more closely their CIs and how they are positioned relative to each other. These comparisons may be useful if one were to be concerned with particular persons' latent score comparisons.[3] (This procedure is obviously best used if prior to the IRT analysis it is known which two or more subjects are to be compared in this way on the latent dimension of concern.)

6.7   Scoring of studied persons

A nearly routine goal in many behavioral measurement and testing settings is the generation of a "score" for each examined subject in a studied group. This score is typically assigned to him or her at the end of an analytic or modeling session on a dataset resulting from an empirical investigation. It is expected from this individual score to reflect as well as possible the degree to which that person possesses the latent trait or ability, θ, that is (presumably) evaluated by a given item set or measuring instrument used in the study. Within the IRT framework, as indicated before, based on the responses provided by each subject in the sample, a latent trait or ability score, $\widehat{\theta}$, is estimated (predicted) for him or her following the methods used earlier in this chapter (section 6.6) and discussed in more detail in the next chapter. Hence, for each examined person, the value $\widehat{\theta}$ can be considered a resulting score that represents an approximation to their unknown trait or ability level θ. The quality of this approximation will depend obviously on the degree to which the used IRT model is an adequate means for description and explanation of the analyzed data.

While this person score $\widehat{\theta}$ is readily determined with contemporary software implementing those estimation methods, the individual value of $\widehat{\theta}$ is not necessarily easily interpretable because it is expressed in what may be seen as an abstract metric. Specifically, in many empirical investigations, it may be difficult for scholars using IRT to communicate to applied researchers and practitioners the meaning of particular noninteger scores or negative estimates that are in a sense as frequent as positive scores

---

3. When the individual observations (subject response sets on the studied items Y_1 through Y_k) are not a priori ranked in some substantively meaningful way, one may wish to consider carrying out the following activity prior to individual trait or ability level estimation (prediction). First, compute the number correct score $W = Y_1 + \cdots + Y_k$ using the `generate` command, and rank-order the dataset in ascending order on `W` using the command `sort W`. Then, create a new identifier within that rank ordering with the command `generate id2 = _n`, and use it in lieu of `id` in the last Stata command in the main text. This will produce the graph on figure 6.4 but with horizontal axis being number correct, that is, the observed proxy `W` for the underlying trait, construct, or ability. This feature will give the opportunity for comparing subsequently particular individual trait parameter estimates while accounting for their inherent uncertainty (see also Goldstein [2011] and Goldstein and Healy [1995]).

among the resulting $\widehat{\theta}$ estimates. To aid the interpretation of the individual trait or ability level estimates, therefore, one can use linear or nonlinear transformations of $\widehat{\theta}$. In the rest of this section, we consider a popular linear transformation, the so-called T-transformation. We return to this general matter of individual trait or ability scoring in chapter 8 in the context of an important nonlinear transformation as an alternative to the T-transformation.

Whereas there are potentially infinitely many possible linear transformations of $\widehat{\theta}$, the one to the so-called T-score has been often used in empirical, behavioral, and social research. This transformation is defined as follows:

$$T = 50 + 10 \times \widehat{\theta} \tag{6.5}$$

As a particularly attractive feature of this transformation, one may see that practically all transformed scores (under normality) are positive and fall within the interval $[0, 100]$. That is, in effect, all T-scores are nonnegative and none exceed 100. However, this numeric interval $[0, 100]$ is also characterized by the property that it covers all percentage correct scores resulting from an instrument with binary or binary scored items. (The latter traditionally used scores are readily interpreted as the ratio of number of correct scores to number of items, in the percentage metric.) In addition, if noninteger scores are of concern and not desirable in an application, rounding the T-scores in (6.5) to the nearest whole number would provide a set of individual scores that are all integer (see below).

We illustrate this discussion with the LSAT example considered earlier in the chapter. To this end, after obtaining as in section 6.6 the individual trait or ability level estimates $\widehat{\theta}$ using the 1PL model that we selected as discussed in section 6.5, we can readily furnish the T-scores with Stata as follows:

```
. generate T = 50 + 10*Theta
```

To see the results of this activity, we list, say, the first 10 transformed T-scores:

```
. list T in 1/10
```

	T
1.	30.89885
2.	30.89885
3.	30.89885
4.	35.71194
5.	35.71194
6.	35.71194
7.	35.71194
8.	35.71194
9.	35.71194
10.	35.71194

Presenting the transformed T-scores in terms of integer numbers may be desirable in some situations to enhance further if need be their interpretability, in what may be viewed then as more pragmatic terms. (We point out, however, that some differentiability between subjects may be lost thereby, specifically for persons with sufficiently close but nonidentical T-scores.) This presentation is readily accomplished with the Stata function `round()` (for rounding to the nearest whole number):

```
. generate T_round = round(T)
```

For the first 10, say, subjects in the LSAT dataset, the resulting rounded T-scores are then as follows:

```
. list T_round in 1/10
```

	T_round
1.	31
2.	31
3.	31
4.	36
5.	36
6.	36
7.	36
8.	36
9.	36
10.	36

Rounded T-scores, like the initially obtained T-scores, may be readily interpretable in applied empirical settings. This ease of interpretation is based in particular on their earlier indicated feature that they in effect are all contained in the same interval that is covered also by the familiar "percentage correct" scores. Hence, one may say that in a sense the T-score is evaluated in the same metric, that is, on the same "scale", as the familiar percentage correct score.

As mentioned earlier in this section, we revisit this topic of IRT-based individual scoring in chapter 8. In that chapter, we will discuss it further within the context of an important nonlinear transformation, which will capitalize once again on the highly useful connections between classical test theory and IRT.[4]

---

4. An alternative linear transformation of the individual responses, which is currently not widely used, could also be used for scoring purposes. This transformation follows from the fact that the linear combination of the original scores Y_j on the k dichotomous items used, with weights being the item discrimination parameters, that is, the sum $a_1Y_1 + a_2Y_2 + \cdots + a_kY_k$, is a sufficient statistic for the latent trait or ability score θ within the 2PL model, $j = 1, \ldots, k$ (for example, Hambleton and Swaminathan [1985]). That is, upon fitting this IRT model (assumed plausible for the analyzed dataset and preferable to other rival models), we see that this alternative individual score results as, say, $s_i = \widehat{a}_1Y_{1i} + \widehat{a}_2Y_{2i} + \cdots + \widehat{a}_kY_{ki}$, where $\widehat{a}_j$ denotes the estimated discrimination parameter for the jth item ($i = 1, \ldots, n, j = 1, \ldots, k$). This score s_i represents a (statistically) optimal scoring formula for the 2PL model and special cases of it (for example, Lord [1980]).

6.8   Chapter conclusion

In this chapter, which is the first in the book with a distinct empirical and software-related focus, we discussed how Stata can be used to fit logistic models. We commenced with pointing out the ways in which one can read in (access) a dataset, covering several different settings of importance in empirical studies as well as descriptive statistic evaluation. We then fit three logistic models and conducted model selection. We used the popular LRT, the Wald test, and two information criteria—the AIC and BIC. We similarly illustrated the estimation (prediction) of individual trait or ability levels, which represent the studied subjects' positions on the underlying unobserved continuum θ of actual interest. We then showed how the resulting individual estimates (predictions) can be graphically displayed along with their CIs or, along the same lines, with other functions of their standard errors (see also Goldstein [2011] and Goldstein and Healy [1995]). We concluded the chapter with a discussion of a popular linear transformation, the T-transformation, which often considerably aids the interpretation of these individual trait or ability level estimates. In the next chapter, we will attend further to the process of IRT model and parameter estimation, which we have used in the present one for its Stata-based illustrations.

7 Item response theory model fitting and estimation

In this chapter, we will attend further to methods and procedures for fitting item response theory (IRT) models and selecting from two or more models considered for a given dataset in an empirical setting. These methods also allow model parameter estimation and were in part discussed and used in the last chapter. We will thus be concerned with additional details on their rationale and justification in the remainder of this chapter.

7.1 Introduction

Once one chooses an item response model to use for an analyzed dataset, to apply IRT, one needs to estimate its parameters. Based on our discussions in chapters 4 and 5, it may at first appear that this activity could be rather similar to that when a regression analysis model and in particular a logistic regression model is used (compare, for example, Cai and Thissen [2015]). However, there is an important difference between this modeling framework on the one hand and IRT on the other. This key difference lies in the fact that in regression analysis and logistic regression, the predictors are observed, that is, recorded individual observations are available for them. However, in IRT, the individual ability scores on the θ dimension, which could be viewed conceptually as predictor scores (see chapters 1, 4, and 5), are not observed. Thus, whenever IRT modeling is used, one generally considers as unknowns both the parameters of the model used (item parameters) and the individual person trait, construct, or ability scores (individual ability levels). Based on this consideration, these individual scores are sometimes also called trait or ability parameters.

Parameter estimation in IRT is currently predominantly conducted using the popular maximum likelihood (ML) method (compare van der Linden [2016b]; for Bayesian-statistics approaches, which are increasingly more often used, see, for example, Cai and Thissen [2015]). This is in part due to its features of yielding with large samples (that is, asymptotically) unbiased, consistent, normal, and efficient estimators (for example, Casella and Berger [2002]). In addition, ML estimators possess the highly useful property of invariance that renders ML estimators for considered functions of parameters by merely substituting their ML estimators in those functions. Moreover, ML estimators (where unique) are functions of sufficient statistics when existing, that is, statistics that contain all information available in a studied dataset with respect to the unknown parameters of interest to estimate (for example, Roussas [1997]). Therefore, in this chapter we discuss the rationale and process of obtaining ML estimates in applications of IRT. To this end, we start with the concept of likelihood (likelihood function) for individual subjects and then generalize the discussion.

7.2 Person likelihood function for a given item set

ML estimates result from maximizing the data likelihood with respect to the unknown parameters, which is referred to as the likelihood function, often called simply "likelihood" (for example, Agresti [2013]). In the settings of relevance to this book, the likelihood represents the probability of the observed data considered as a function of the unknown parameters (Casella and Berger 2002). For our purposes in this chapter, it is instructive to begin the likelihood function discussion with the single person case, and in the setting of binary or binary scored items, that is assumed until chapter 11 (where polytomous items and IRT models for them will be discussed).

For a given person under consideration, the likelihood function is based on his or her data or observed scores denoted $y_1, y_2, \ldots, y_k$ on the k items administered to him or her ($k > 1$). The scores y_j represent the individual realizations of the binary random variables associated with each of these items, designated $Y_1, Y_2, \ldots, Y_k$, respectively ($j = 1, \ldots, k$). Once we fix this person, we in effect fix his or her trait or ability level, that is, θ. We can thus use the local independence assumption that we made throughout the book, as is typical in current IRT utilizations (for example, see van der Linden [2016b]; see also chapter 5). Thus, the associated likelihood function is as follows,

$$P(Y_1, Y_2, \ldots, Y_k | \theta) = P(Y_1 | \theta) P(Y_2 | \theta) \ldots P(Y_k | \theta) = p_1 p_2 \ldots p_k \qquad (7.1)$$

where P symbolizes generically probability (conditional probability) and $p_1, p_2, \ldots, p_k$ designate the conditional probabilities for the individual responses on the respective items for a given θ. Because each Y_j is binary, we can write these individual probabilities, once the data are given, also as

$$p_j = P_j^{y_j} Q_j^{(1-y_j)} \qquad (7.2)$$

where P_j is the probability of the response denoted 1 given θ and $Q_j = 1 - P_j$ ($j = 1, \ldots, k$). The reason for the validity of (7.2) is that in case $y_j = 1$, only the first factor in its right-hand side matters and remains as P_j (because $Q_j^{(1-y_j)} = Q^0 = 1$). Conversely, if $y_j = 0$, only the term Q_j is of relevance and remains instead ($j = 1, \ldots, k$).

Based on (7.2), the right-hand side of (7.1) can then be rewritten as

$$P(y_1, y_2, \ldots, y_k | \theta) = \prod_{j=1}^{k} P_j^{y_j} Q_j^{1-y_j} \qquad (7.3)$$

where the symbol $\prod$ is used to denote multiplication (across the range of the subindex j indicated, that is, from $j = 1$ through $j = k$). We stress that when the data of the person of interest are given on the k items of concern, the right-hand side of (7.3) is a function of their ability θ (Casella and Berger 2002). Hence, his or her likelihood function is the right-hand side of (7.3). To make this reference to the likelihood function explicit, we use the notation $L(\mathbf{y}|\theta)$ for that side of (7.3). Denoting by $\mathbf{y}$ the vector of data of this person on the items, his or her likelihood function is then

$$L(\mathbf{y}|\theta) = \prod_{j=1}^{k} P_j^{y_j} Q_j^{1-y_j} \tag{7.4}$$

For example, if Johnny Smith (for simplicity referred to as JS below) has solved correctly only the first, second, and fourth tasks on an algebra test consisting of five items, say, then in the notation introduced above, his likelihood function is

$$L(1, 1, 0, 1, 0|\theta) = P_1 P_2 Q_3 P_4 Q_5$$

where θ is his unknown algebra ability level, which is evaluated by the test.

We stress that although this is not explicated in the above equations, the likelihood function of a person—once a set of items is given—depends strictly speaking both on his or her ability level θ and on the parameters of those items. These parameters are introduced into the likelihood function once the model is settled on, which will be used in a IRT modeling session (see also next section).

To see this dependency in more detail in an illustrative example, given an IRT model, let us focus on the above five-item algebra test and the data for JS on it. Assuming that the one-parameter logistic (1PL) model, say, is an adequate model for his data, the expressions for the above probabilities P and Q associated with his likelihood function are as follows,

$$P_j = \exp\{a(\theta - b_j)\}/[1 + \exp\{a(\theta - b_j)\}], \text{ and}$$
$$Q_j = 1/[1 + \exp\{a(\theta - b_j)\}]$$

$j = 1, \ldots, 5$. Hence, the likelihood function associated with the data is

$$\begin{aligned}
L(\text{JS}' \text{ data}|\theta) = {}& \exp\{a(\theta - b_1)\}/[1 + \exp\{a(\theta - b_1)\}] \\
& \times \exp\{a(\theta - b_2)\}/[1 + \exp\{a(\theta - b_2)\}] \\
& \times 1/[1 + \exp\{a(\theta - b_3)\}] \\
& \times \exp\{a(\theta - b_4)\}/[1 + \exp\{a(\theta - b_4)\}] \\
& \times 1/[1 + \exp\{a(\theta - b_5)\}] \tag{7.5}
\end{aligned}$$

From (7.5), we readily see that the likelihood function for a given person (and model used) depends on the item discrimination parameters and the item difficulty parameters. (The function obviously also depends on multiple item discrimination parameters and possible pseudoguessing parameters if the more general two-parameter logistic (2PL) or three-parameter logistic (3PL) model is used.) In addition, and no less importantly, this likelihood is also a function of the individual ability level, namely, the unknown θ-value of JS in this example.

At this moment, we quickly realize that we are facing a rather complex estimation problem. Specifically, we are given only five numbers as observations, namely, the responses of JS on the five items that have been administered to him. However, there are actually up to 16 unknown item parameters in the most general popular IRT model, the 3PL model (namely, 3 parameters per item plus his θ value). Even if we choose the simplest logistic model, which is the Rasch model, we are still faced with the task of estimating seven parameters for him. These are the item discrimination and five-item difficulty parameters and his underlying trait or ability level θ.

To make progress step by step, unless otherwise stated in the rest of this section 7.2, we will assume that the item parameters are known. This may be a reasonable assumption when prior to a particular analysis, the items were administered to a large and representative sample from the studied (subject) population, and their parameters were estimated from the resulting data. (This is an issue that we will attend to later in the chapter.) The latter procedure is often called "item calibration", deals with sets of items to be used subsequently for certain purposes (see, for example, chapter 9), and furnishes their estimated parameters and related quantities and functions. These estimates can frequently be reasonably treated as known in ensuing analyses using these items, especially when their standard errors (SEs) are small (see below). We also mention that the item parameter estimation can be accomplished, for instance, using a "modified" ML method that is referred to as marginal maximum likelihood (MML), which is discussed later in this chapter.

7.2.1   Likelihood reexpression in log likelihood

Returning now to our discussion of individual ability estimation, using (7.4), we note first that its right-hand side is typically a rather small number (at any given value of the unknown parameter it depends on). How small it will be in a particular setting depends on the number of items and their pertinent probabilities of relevance. (As indicated earlier, the latter are determined themselves by the model, item parameters, and ability level of the considered subject). That is, in general, the right-hand side of (7.4) can be a very small number. However, it is quite difficult to work with such small numbers in their "raw" metric. Yet it is much easier and convenient to work instead with their logarithms—at times referred to as "logs"—that are considerably larger in magnitude (in absolute value terms). Thus, as in other applications of the ML estimation method, we will work with the logarithm of the likelihood function that is here as follows [see (7.4)]:

$$l(\mathbf{y}|\theta) = \ln\{L(\mathbf{y}|\theta)\} = \sum_{j=1}^{k}\{y_j \ln P_j + (1 - y_j)\ln Q_j\} \tag{7.6}$$

In (7.6), which is the general log-likelihood expression of relevance for person trait or ability level estimation, $\ln(\cdot)$ denotes as earlier the natural logarithm function, that is, the logarithm with base $e = 2.718\ldots$. We stress that once the data for a single examined person are provided, $l(\mathbf{y}|\theta)$ in (7.6) is a function of θ. Hence, when interested in the log likelihood as such a function, we can write it simply as $l(\theta)$. (Recall from our earlier discussion that we assumed the item parameters as known.)

7.2.2 Maximum likelihood estimation of trait or ability level for a given person

With the preceding discussion in mind, to find the ML estimate of a given person's ability level θ, we maximize the log-likelihood function $l(\theta)$ using "first principles" of calculus (for example, Zorich [2016]). Accordingly, we take its first derivative, set it at 0, and then solve the resulting equation for θ. (The general ML theory ensures that under certain regularity conditions the solution of this equation, called "normal equation", is a maximum of the likelihood; see also below.) However, this equation does not typically admit a closed form solution. Hence, it needs to be solved using numerical optimization methods that are implemented in widely circulated statistical software such as Stata, which we already saw in action in the last chapter. The numerical optimization methods often capitalize on an appropriate version of the popular Newton–Raphson procedure, with alternative approaches also being frequently used (see Cai and Thissen [2015]; see also van der Linden [2016b]).

Relatedly, it should be noted that in general the likelihood function might not have a finite value at its maximizer, denoted in this book as $\widehat{\theta}$. That is, its maximum may be infinite (indefinite). This occurs when a person correctly answers all items in a given set (test), or none of them. For such persons and response patterns, the ML estimates of their ability levels do not exist. The reason is that they cannot be determined with any degree of precision based on their responses, because the latter are lacking enough information for this to be possible. We address this issue in a following subsection. Similarly, when a person responds correctly to some "harder" items but incorrectly to some "easier" items, for instance, in a 3PL model, the ML estimate of his or her ability may not exist. We also attend to this problem later in the chapter.

The general ML estimation theory implies an important fact that we will use repeatedly throughout the rest of this book (compare Casella and Berger [2002]; see also next subsection). Specifically, for a given person, as k increases indefinitely (that is, a unidimensional set of more and more items are administered to him or her that follow an adequate, fit IRT model), the SE associated with the ML estimator of his or her ability level can be obtained. Denoting this SE as $SE(\widehat{\theta})$, it follows from the ML theory that

$$\mathrm{SE}\left(\widehat{\theta}\right) = 1/\sqrt{I(\theta)} \tag{7.7}$$

holds then, where $I(\theta)$ is the so-called information matrix (information function) for the setting considered here, which is defined formally below. Obviously, the right-hand side of (7.7) cannot be computed directly, because the person's ability level θ is not known. However, one can substitute in it his or her estimated ability level, $\widehat{\theta}$. Then the (estimated) SE associated with the ML estimate of that person's ability level, denoted in the rest of this chapter for convenience simply as SE, becomes

$$\text{SE} = 1/\sqrt{I\left(\widehat{\theta}\right)} \tag{7.8}$$

The ML theory also implies, as mentioned earlier, large-sample normality of the ML estimator, under pertinent conditions, referred to as "regularity conditions" (for example, Casella and Berger [2002]). Thus, a 95% confidence interval (CI) for a given person's ability level is obtained as

$$\left(\widehat{\theta} - 1.96\ \text{SE}, \widehat{\theta} + 1.96\ \text{SE}\right) \tag{7.9}$$

[We note that we would use as a multiplier the number 1.64 for a 90% CI and the number 2.58 for a 99% CI, instead of 1.96 in (7.9); see Agresti and Finlay (2009)]. We stress that because the person under consideration is fixed here, the term "large sample" used in this paragraph means only increasing the number of items administered to him or her that follow the fit IRT model as indicated earlier. [In this sense, strictly speaking, (7.9) is to be considered an approximate 95% CI for the individual trait or ability level, θ.]

From (7.7) and (7.9), we see that both the SE and width of the CI for the estimated individual trait or ability level depend on the actual, unknown level θ. Hence, both the SE (information function) and width of associated CIs will have in general different values at different ability levels (estimates). That is, the SE of the individual trait or ability levels and the length of the associated CI differ across θ values.

7.2.3 A brief visit to the general maximum likelihood theory

We are now ready to attend to the general ML theory in the context of IRT and item response modeling (IRM), specifically for the case of ML estimation of individual trait or ability levels. (In this subsection, as will be recalled, we have fixed a person under consideration.) As indicated earlier in the book, under certain regularity conditions and with large samples (that is, asymptotically), the ML estimates are consistent, unbiased, normal, and efficient (that is, with smallest variance); furthermore, the ML estimates possess the invariance property and, if unique, are functions of sufficient statistics when they exist (for example, Casella and Berger [2002]). That is, no further information in the data is necessary to obtain the ML estimates beyond that needed for furnishing the pertinent sufficient statistics.

Moreover, and of special relevance in this section, for the ML estimate of ability resulting from an adequate model, the following holds (for example, Casella and Berger [2002])

$$\widehat{\theta} \sim^a N[\theta, \{I(\theta)\}^{-1}]$$

where "$\sim^a$" is used to symbolize large-sample (asymptotic) distribution (compare chapter 2). In other words, the ML estimate of individual ability is asymptotically normal (that is, for a large number of unidimensional items), with a mean being the true ability level, θ, of the examined person, where $I(\theta)$ is the information function. In the current setting, this information function is given by the following expression (see specifically its last part):

$$I(\theta) = -E\left(\partial^2 \ln L/\partial\theta^2\right) = \sum_{j=1}^{k}\left\{\frac{d}{d\theta}P_j(\theta)\right\}^2 / \{P_j(\theta)Q_j(\theta)\} \qquad (7.10)$$

In (7.10), the symbol $E(\cdot)$ is used to denote expectation (of the expression within brackets), and within the first and second pair of brackets, partial derivatives (in general) with respect to θ are used, respectively (see, for example, Baker and Kim [2004]; see also next chapter).

From (7.10), we see that the information function depends on the model settled on (because the model determines the item characteristic curve involved), in addition to the ability level in question, θ. Specifically, with respect to the SE associated with a particular subject's ability level, from this discussion, it follows that its estimate is

$$\left\{\text{SE}\left(\widehat{\theta}\right)\right\}^2 = V\left(\widehat{\theta}|\theta\right) = \left\{I\left(\widehat{\theta}\right)\right\}^{-1}$$

In (7.10), $V(\cdot|\cdot)$ denotes estimated variance (which depends on θ as mentioned earlier), and $\{\text{SE}(\widehat{\theta})\}^2$ designates the square of the estimated SE of the ability estimate, $\widehat{\theta}$.

Therefore, the width of the CI (7.9) for the individual trait or ability level θ (that is, the precision of its estimation) depends inversely on the information function value involved. This is because the latter determines the SE appearing there. Specifically, the larger or smaller the value of the information function is, the smaller or larger the width of the CI is; hence, the more or less precise is the estimation of the ability level θ. (We keep in mind that the information function itself, as mentioned, is a function of θ.) We elaborate further on these and related matters in the next chapter.[1]

---

1. The incorrect view of (an individual trait or ability estimate or prediction) $\widehat{\theta}$ as an observed quantity seems still to be surprisingly widespread among some circles of IRT users and promoters. In an IRT context, only the responses of the n studied subjects on the k administered items to them are observed quantities (observed scores). (These are the individual realizations of the earlier introduced random variables Y_{ji}, which symbolize the answer of the ith person to the jth item; $i = 1, \ldots, n, j = 1, \ldots, k$; see, for instance, chapter 3. In particular, the only observed are the quantities or scores denoted y_j in section 7.2 of the present chapter.) Indeed, as elaborated in this chapter, for a given person, his or her $\widehat{\theta}$ results from an involved estimation process and cannot be logically viewed as an observed quantity (also because $\widehat{\theta}$ is part of the end result of an IRT modeling session or application). An even simpler argument would suffice to realize this fact as well: because $\widehat{\theta}$ is associated with a SE, namely, the quantity defined in (7.8) (see also earlier presented output sections from used empirical examples, in particular in chapter 6), this estimate or estimator $\widehat{\theta}$ cannot possibly be an observed quantity—obviously, because no observed quantity can be, or need be, associated with an SE(!).

7.2.4   What if (meaningful) maximum likelihood estimates do not exist?

This estimation problem, indicated in the preceding subsection for persons solving all or none of the items correctly, can be handled using a Bayesian estimation procedure. The latter is based on the so-called Bayes theorem (for example, Raykov and Marcoulides [2013]). Accordingly, the probability of an event, say, A, given another event, say, B, with both having positive probabilities, is

$$P(A|B) \sim P(B|A) \times P(A) \tag{7.11}$$

where "$\sim$" stands for "proportional". An analog of (7.11) exists for probability density functions (PDFs), denoted by h next, considering the parameter θ as a random variable following a certain distribution [see, for example, Johnson and Sinharay [2016]; see also (7.4)]:

$$h(\theta|\mathbf{y}) \sim h(\theta)L(\mathbf{y}|\theta) \tag{7.12}$$

That is, the PDF of θ once updated given the data, called posterior PDF, is proportional to the product of an assumed a priori PDF for θ (called prior) with the likelihood function.

What (7.12) now suggests is the following sensible estimation approach. If one knew or was willing to assume initially a certain distribution for the ability level, denoted $h(\theta)$ and referred to as the prior distribution of θ, then one could obviously use it—via (7.12)—in the process of estimation of the ability level for the person in question. This estimation is achieved via "updating" the prior distribution by using the likelihood function, $L(\mathbf{y}|\theta)$ [see, for example, (7.4)], to obtain the revised or updated individual ability distribution from (7.12), namely, $h(\theta|\mathbf{y})$. The latter is typically referred to as the posterior distribution of θ and is of special relevance in applications of Bayesian statistics.

Using a "diffuse" prior, for example, the normal distribution with a fairly large variance, one can then use the mode of the posterior distribution or its mean as an individual ability estimate for any person with nonexistent ML estimate (see above in this section). This approach is also built into widely circulated software and is readily used in empirical settings (see, for example, section 6.6 in chapter 6.)

7.3   Estimation of item parameters

Our discussion in the last section assumed the item parameters as known. However, they are usually not known, at least at the beginning of an IRT application session. To deal with this issue, first let us recall how we proceeded in the last section to estimate the individual ability level of a given person via ML—we administered a large number of unidimensional items (following a fit IRT model) and obtained the likelihood function associated with his or her responses.

The same general idea can be used for item parameter estimation (compare, for example, Hambleton, Swaminathan, and Rogers [1991]). To this end, we can assume that we knew the ability levels (see next section for what can be done otherwise). To estimate the parameters of a given item then, let us administer it to a large number of persons (say, n subjects). Once we have their binary (binary scored) results on the item, for a given logistic IRT model of concern, we obtain similarly to the previous developments in this chapter the following data likelihood (for the most general, 3PL model with a discrimination, difficulty, and pseudoguessing parameter):

$$L(\mathbf{y}|\theta, a, b, g) = \prod_{i=1}^{n} P_i^{y_i} Q_i^{1-y_i} \tag{7.13}$$

In (7.13), $\mathbf{y}$ denotes the set of these n persons' responses on the item in question, and the P_i and Q_i are the pertinent probabilities of correct and incorrect responses, respectively, as functions of associated parameters according to the model [$i = 1, \ldots, n$; see (5.30) in chapter 5]. In obtaining the likelihood function (7.13), we emphasize that one does not need to make the local independence assumption, which, as will be recalled, we had to use to obtain the likelihood function for a given person and set of items administered to him or her in section 7.2. The reason is that, as is the case in conventional or standard statistical methods, persons are assumed to be responding to the item in question independently of each other. This classical statistical assumption is made throughout the book. (The assumption is relaxed in two-level or multilevel modeling settings that are not considered in the remainder of this book but were mentioned toward the end of chapter 1; see also section 4.4 in chapter 4 and epilogue.)

When the ability levels for the subjects in an available sample are known, the only difference from the case dealt with in the preceding subsection is that we have now in general (up to) a three-dimensional likelihood function to maximize. This differs from the situation in the last section, in that we had there a unidimensional, that is, single-parameter likelihood function to maximize. Either way, we can use the same general numerical approach—namely, numerical optimization—to solve the "normal" equations obtained by setting equal to zero the three partial derivatives of the likelihood function (7.13) in the general case. Use, for instance, of the general Newton–Raphson procedure can be made then, as indicated earlier in this chapter. By applying this procedure for each item, we obtain ML estimates for all its parameters and then those for all considered items.

7.3.1 Standard errors of item parameter estimates

Earlier in this chapter, we introduced the concept of information function. We also noted that it is a fundamental notion in the ML theory that is directly used when using ML in IRM and IRT applications. In particular, as indicated before, when the ML estimate of an ability level was obtained (under regularity conditions as mentioned), the reciprocal of the square rooted information function gave the associated SE [see, for example, (7.8)].

Along the same line of reasoning, that is, application of the general ML theory, when ML estimates of the item parameters are obtained, for each item, there is a pertinent information matrix associated with its up to three parameters. To furnish the SEs of the item parameters, one can invert that matrix and take the square roots of its main diagonal elements. To be more specific, for a given item (say, the jth), this matrix is in the general case (compare, for example, Baker and Kim [2004]; $j = 1, \ldots, k$):

$$I_j(a_j, b_j, g_j) = \begin{bmatrix} i(j,a) & i(j,ab) & i(j,ag) \\ & i(j,b) & i(j,bg) \\ & & i(j,g) \end{bmatrix} \tag{7.14}$$

In (7.14), $i(.,.)$ is used to denote matrix element, and the elements on the main diagonal pertain to the individual item parameters (that is, its a, b, and g parameters) and the off-diagonal elements to pairs of item parameters. [This is obviously a symmetric matrix, and so only its main diagonal elements and those above it are presented in (7.14).] Widely circulated software can estimate (7.14) for each item and invert that matrix to furnish the covariance matrix of the parameters of a given item. From that matrix, correspondingly, the SEs of the item parameters are obtained as the square roots of its diagonal elements. The general expressions for the elements of the information matrix for a given item, as functions of ability levels, item parameters, and model-implied probabilities of correct response, are provided, for instance, in Baker and Kim (2004). (As for MML, discussed below, we mention that simple expressions for the covariance matrix of the resulting estimates are not available. However, they can be well approximated numerically, as implemented also in widely circulated software; see, for example, Mislevy and Bock [1984].)

7.4   Estimation of item and ability parameters

In a typical behavioral, educational, or social study, neither the ability levels nor the item parameters are (initially) known. The question that naturally arises is how to deal with that situation, which is realistic for very many cases in empirical research. We can begin to approach this serious estimation problem by first looking at the likelihood function for n persons' responses on k given items:

$$L(\mathbf{y}|\boldsymbol{\theta}, \mathbf{a}, \mathbf{b}, \mathbf{g}) = \prod_{i=1}^{n} \prod_{j=1}^{k} P_{ij}^{y_{ij}} Q_{ij}^{1-y_{ij}} \tag{7.15}$$

In (7.15), y_{ij} is the response of the ith person on the jth item, P_{ij} and Q_{ij} are the associated (model-specific) probabilities for "correct" and "incorrect" response, respectively. Further, in that equation, bolding is used to denote a vector or set of parameters. Specifically, the n individual trait or ability levels or parameters are placed in the vector $\boldsymbol{\theta}$, the k item discrimination parameters in $\mathbf{a}$, their difficulty parameters in $\mathbf{b}$, and in $\mathbf{g}$ their pseudoguessing parameters in the most general, 3PL model.

Once the data are given, the right-hand side of (7.15) is a function of up to $3k$ item parameters (in the 3PL model), in addition to the n ability levels. We stress that we need to assume again local independence to be in a position to estimate here the ability levels. There is, however, a new problem that we are facing now. This is the issue of parameter indeterminacy or lack of identification, which we indicated earlier in the book.

To be more concrete, let us suppose we replaced the ability levels with $\gamma = \alpha\theta + \beta$ (with an arbitrary nonzero α) and for each item its difficulty parameter with $\alpha b + \beta$. If we then also replaced its discrimination parameter by a/α, as a result, the probability of correct response on a given item will remain unchanged (note that the following equation holds regardless of whether there is guessing on the item):

$$P(\theta) = P(\gamma)$$

As a consequence of this lack of unique parameter estimability based on the data, that is, lack of parameter identification, the likelihood function will not have a unique maximum. Hence, for a given dataset, we will not be able to come up with a unique set of parameter and trait or ability level estimates. That is, all of our IRT-related efforts will be in vain. So how can we deal with this serious identification issue?

This problem is rooted in the fact that we are fundamentally interested in having also the unknown individual trait or ability levels as instrumental parts, or parameters, of our models. However, the corresponding unknown trait or ability is not directly measured. Therefore, there is no natural origin and unit of measurement for it, as there would be if that trait was directly measured, that is, if observations and measurements of θ itself were (recorded or made) available for the studied persons. The issue at hand is, however, not unique to IRT. In fact, the same type of problem emerges when we involve a latent variable in a model, for example, a factor analysis or structural equation model (for example, Raykov and Marcoulides [2006]). The source of the problem here is that the latent variables, being unobserved or unobservable, do not have an inherent or natural metric (scale). Thus, to compensate for that, we need to introduce or fix such a metric in our modeling or analytic session.

However, there is a relatively direct way of resolving this problem, when dealing with a single group of persons (and a single assessment or measurement occasion as in this book). This approach, as mentioned earlier, consists of setting the mean and variance of the ability level variable, θ, to 0 and 1, respectively. This pair of constraints are actually invoked automatically in most IRT software when an IRT model is fit to data from a single group (population). Once these constraints are introduced, one could for instance use the so-called joint maximum likelihood (JML) method to maximize the likelihood function (7.15). An application of JML for this purpose would amount, however, to a multistep process (compare, for example, Hambleton, Swaminathan, and Rogers [1991]). In the first step, one would need to choose appropriate initial values for the ability levels or parameters. Software default routines, based on pertinent algorithms, will perform well most of the time, but the logarithms of the ratios of number of correct to number of incorrect responses per person will often likely do well, too. As a next step, one would want to standardize these initial values so that they have a mean of 0 and variance of 1. Then, treating these standardized initial values as known, one would

estimate the item parameters, as discussed earlier in this chapter. Subsequently, treating the so-obtained item parameters as known, one could estimate the individual ability
levels, as we discussed in a previous section of the chapter. Using the last three steps
repeatedly (iteratively), until the resulting estimates are essentially identical across consecutive iterations, will render in the end JML estimates of all parameters of interest.
This approach to ML estimation has its origins for the logistic models in the pioneering
work by Birnbaum in the 1960s and was quite popular a few decades ago. In fact, it may
be considered a version of a more general EM algorithm that has been used oftentimes
since then for IRM in empirical settings (compare, for example, Cai and Thissen [2015]).

While conceptually clear, and in fact presenting a solid basis for more sophisticated
methods of IRT model and parameter estimation, this JML procedure has several important problems (for example, Hambleton and Swaminathan [1985]). One is that for
subjects having solved correctly all or none of the items, JML cannot produce ability
estimates. Similarly, JML cannot produce estimates for items that are solved correctly
by all subjects, or solved incorrectly by all persons. Further, this estimation method
does not produce consistent estimates of item and ability parameters, particularly in
two- and three-parameter models. This is due to the problem of so-called incidental
parameters, which results when the number of parameters increases with increasing
number of subjects, or sample size (for example, Baker and Kim [2004]). Last but not
least, JML is prone in general to convergence failures, especially with three-parameter
models.

To resolve these problems, we can pursue two avenues (for example, van der Linden
[2016b]). In the first approach, prior distributions are placed on the item and ability
parameters, leading to Bayesian estimates of all of them. This is a closely related
approach to the one mentioned earlier in this chapter for obtaining estimates of trait or
ability levels for persons with all correct or all wrong responses on the used items. In
the second approach, referred to as MML, a distribution is first assumed or placed on the
subjects' ability levels, such as the normal distribution. Based on it, the expectation is
taken of the likelihood function in (7.15). This process is also referred to as "integrating
out" the ability parameters from the likelihood function. (An alternative reference
is "marginalization" of the likelihood.) The process leads to the so-called marginal
likelihood (that is, independent of the subjects' θ's), which is then maximized. The
resulting item parameter estimates are called MML estimates. They are consistent as
the number of examined subjects (sample size n) increases indefinitely. This approach
can also be seen as another version of an EM algorithm application and is described in
detail in Bock and Aitkin (1981).

A potential issue with MML is the computational burden from the integration that
needs to be conducted. Further, it is necessary to have a very good approximation
or assumption for the distribution of the individual ability levels in the studied population. The number of examined persons typically needs to be large then, preferably
in the thousands (compare Hambleton, Swaminathan, and Rogers [1991]). Once MML
estimates are obtained, however, for the item parameters, the latter may be treated as
subsequently known. Then individual ability levels are estimated as outlined earlier in

this chapter, for example, using the ML or Bayesian estimation procedures. However, numerical problems with MML can occur with three-parameter models, specifically when there is no sufficiently good information in the sample to allow satisfactory estimation of the pseudoguessing parameters. The Bayesian estimation approach can deal with this problem by placing a prior distribution on these parameters and proceeding using correspondingly Bayesian methods (for example, Cai and Thissen [2015]).

7.5   Testing and selection of nested item response theory models

As indicated in chapter 6 and earlier in the book, the three logistic models for binary or binary scored items that are of concern in this book—the 1PL, 2PL, and 3PL models—build a sequence of nested models. Specifically, the 1PL model is nested in the 2PL model, and the latter in turn is nested in the 3PL model. By the obvious transitivity of the nested model relation, one can readily see that the 1PL model is also nested in the 3PL model. The reason is that the 1PL model is obtained from the 2PL model by setting all item discrimination parameters to be the same (or even equal to 1 in some parameterizations of the 1PL model). Similarly, the 2PL model is obtained from the 3PL model by fixing to 0 the pseudoguessing parameter, g, for each item (with such a parameter). Lastly, the 1PL model is obtained from the 3PL model by setting the discrimination parameters the same across items and the pseudoguessing parameters to be 0 for all items.

According to the ML theory, when the ML method of parameter estimation is applicable under pertinent regularity conditions (for example, Casella and Berger [2002]), one can test any pair of the above nested models against each other. That is, one can test whether the 1PL model would be preferable to the 2PL model and to the 3PL model if need be. Alternatively, one can also test if the 2PL model would be preferable to the 3PL model. To this end, one can use the likelihood-ratio test (LRT). As discussed in section 6.3, the LRT approach is based on the fact that with large samples (-2) times the difference in the maximized log likelihoods of the nested model and the more general model follows a chi-squared distribution under the null hypothesis stipulating that the nesting restriction is valid in the studied population [see (6.2)]. We also demonstrated in chapter 6 that the LRT is readily applicable with Stata. This test is recommended to use when the 1PL model is considered as a rival means of data description and explanation relative to the 2PL model or 3PL model if need be. Similarly, the LRT is recommended when one is interested in testing whether the 2PL model is a preferable means of data description and explanation relative to a 3PL model. In addition, the popular Akaike information and Bayesian information criteria can be used (see chapter 6 for their definitions). Thereby, the preferred model in the end possesses the smallest of them across all considered models fit to the same dataset (for example, Raftery [1995]).

7.6   Item response model fitting and estimation with missing data

Incomplete datasets are oftentimes encountered in empirical behavioral, educational, and social studies. Frequently, there is little if anything that a researcher can do to preclude their occurrence. The topic of missing data has attracted an enormous amount of interest by methodologists and substantive researchers over the past several decades (for example, Little and Rubin [2002]). Different mechanisms of missing data have been identified thereby. They have important consequences for how to manage the exceedingly complicated missing data problem (for example, Enders [2010]). A "popular" mechanism of missing data is missing at random (MAR). MAR is currently oftentimes assumed in empirical behavioral and social research. Under MAR, the probability of missingness is not related to the actually unobserved (missing or unavailable) values but may instead be related to observed data. Like the well-known missing completely at random mechanism, MAR is not testable statistically (for example, Raykov [2011]). However, based on substantive considerations and study design and execution details, a researcher may be in a position to argue in favor of MAR. This would in general be possible, for instance, when data are missing by design (for example, Graham [2009]).

When MAR holds, maximizing the likelihood of the observed data yields ML estimates that are not (asymptotically) biased (for example, Enders [2010]). Thereby, all one needs to do to obtain the ML estimates is maximize that likelihood obtained by accumulating the individual contributions from the studied persons in the sample that are based on their present data only (Little and Rubin 2002). This approach is often referred to as "full information" maximum likelihood (FIML). This reference is typically used when one wants to emphasize that all available data from all studied subjects are used from all observed variables in a model of interest. (We stress that FIML is merely an application of ML in the presence of missing data; hence, a reference to this method simply as ML estimation for incomplete datasets would suffice just as well.) No assumption of normality of the observed data is needed for an FIML application when MAR holds. Further, the manifest variables need not even be continuous but could for instance be binary (or binary scored) as in most chapters of this book. That is, FIML is also directly applicable with discrete items, that is, for fitting IRT models, as long as data are MAR. This is in fact a method implemented in a number of IRT programs, including Stata's command `irt`, for analysis and modeling of incomplete datasets. To use it, under MAR, one proceeds at the software interface level in the "same way" as one would if there were no missing data, after first declaring, of course, the missing data to the software (that is, indicating their flags in the analyzed dataset). Interpretation of the obtained results thereby is as in the case of complete (that is, no missing) data.

We illustrate this approach using adapted data from a short survey with $k = 5$ questions of $n = 1165$ young adults from a large city on their attitude toward a decision by the city council. The data are scored 0 or 1 correspondingly for "disapprove" or "approve" on each of the 5 questions about particular aspects of the decision. These data are found in the ASCII (text only) dataset `biwmv.dat`, where (-9) denotes a missing value. Upon reading the data (see section 6.1 in chapter 6), we believe the first activity that's advisable to become involved in with any incomplete dataset is declaring the missing values to the software, followed by saving the resulting dataset in Stata format for future use:

```
. infile id q1-q5 using http://www.stata-press.com/data/cirtms/biwmv.dat
(1,165 observations read)
. describe
Contains data
  obs:          1,165
  vars:             6
  size:        27,960
```

variable name	storage type	display format	value label	variable label
id	float	%9.0g		
q1	float	%9.0g		
q2	float	%9.0g		
q3	float	%9.0g		
q4	float	%9.0g		
q5	float	%9.0g		

```
Sorted by:
     Note: Dataset has changed since last saved.
```

We are now ready to declare to the software the missing values, which as mentioned above are flagged by the number (-9) in the original data file we just read in. To this end, we use the following Stata command:

```
. mvdecode _all, mv(-9=.)
          q1: 3 missing values generated
          q2: 2 missing values generated
          q3: 3 missing values generated
          q4: 3 missing values generated
          q5: 2 missing values generated
```

As seen from Stata's response, the software has encoded on our behalf as missing a total of 13 values using the default symbol of ".". (We stress that in Stata, the symbol "." is used to denote a missing value.) We then save this dataset for future use, before commencing our IRT modeling:

```
. save "biwmv_miss.dta", replace
file biwmv_miss.dta saved
```

If interested in examining now whether the 1PL or 2PL model may be preferable as a means of describing this dataset, we can formally proceed as illustrated earlier in the book (see chapter 6; note that the same commands are used next as in the complete data case of relevance in that chapter, with the only difference now that we use the symbols q* as a shorthand for q1-q5). To this end, we start, say, with the more general 2PL model, which we fit and save its results as follows:

```
. irt 2pl q*
```

Fitting fixed-effects model:

```
Iteration 0:    log likelihood = -2880.4462
Iteration 1:    log likelihood = -2863.0185
Iteration 2:    log likelihood = -2862.7954
Iteration 3:    log likelihood = -2862.7953
```

Fitting full model:

```
Iteration 0:    log likelihood = -2854.7233
Iteration 1:    log likelihood = -2841.4327
Iteration 2:    log likelihood =  -2838.638
Iteration 3:    log likelihood = -2838.6107
Iteration 4:    log likelihood = -2838.6106
```

```
Two-parameter logistic model                Number of obs    =      1,165
Log likelihood = -2838.6106
```

	Coef.	Std. Err.	z	P>|z|	[95% Conf. Interval]	
q1						
Discrim	1.072674	.2962353	3.62	0.000	.4920631	1.653284
Diff	-2.955167	.6063222	-4.87	0.000	-4.143537	-1.766797
q2						
Discrim	.9433289	.2737532	3.45	0.001	.4067825	1.479875
Diff	-1.372616	.303868	-4.52	0.000	-1.968186	-.7770456
q3						
Discrim	1.054578	.3309872	3.19	0.001	.4058552	1.703301
Diff	-.4606289	.1208581	-3.81	0.000	-.6975064	-.2237513
q4						
Discrim	.1356038	.1257478	1.08	0.281	-.1108574	.382065
Diff	-7.5494	6.962297	-1.08	0.278	-21.19525	6.096451
q5						
Discrim	.1637673	.1642112	1.00	0.319	-.1580807	.4856152
Diff	-10.87549	10.80307	-1.01	0.314	-32.04911	10.29813

```
. estimate store m2pl
```

We are now ready to move on to the 1PL model and the application of the LRT, which we achieve in this way:

```
. irt 1pl q*
```

Fitting fixed-effects model:

```
Iteration 0:    log likelihood = -2880.4462
Iteration 1:    log likelihood = -2863.0185
Iteration 2:    log likelihood = -2862.7954
Iteration 3:    log likelihood = -2862.7953
```

Fitting full model:

```
Iteration 0:    log likelihood = -2852.6226
Iteration 1:    log likelihood = -2849.4345
Iteration 2:    log likelihood = -2848.6643
Iteration 3:    log likelihood = -2848.6632
Iteration 4:    log likelihood = -2848.6632
```

```
One-parameter logistic model                    Number of obs   =      1,165
Log likelihood = -2848.6632
```

		Coef.	Std. Err.	z	P>|z|	[95% Conf.	Interval]
	Discrim	.6006169	.0686862	8.74	0.000	.4659944	.7352393
q1							
	Diff	-4.756098	.5301664	-8.97	0.000	-5.795205	-3.716991
q2							
	Diff	-1.973893	.2305084	-8.56	0.000	-2.425681	-1.522105
q3							
	Diff	-.7155907	.1288112	-5.56	0.000	-.9680561	-.4631253
q4							
	Diff	-1.828333	.216884	-8.43	0.000	-2.253418	-1.403248
q5							
	Diff	-3.152789	.3490727	-9.03	0.000	-3.836958	-2.468619

```
. lrtest m2pl

Likelihood-ratio test                            LR chi2(4)  =      20.11
(Assumption: . nested in m2pl)                   Prob > chi2 =     0.0005
```

Because the LRT is significant, we reject the tested hypothesis of equality of the item discrimination parameters that is a characteristic feature of the 1PL model. This finding suggests that the 2PL model is preferable to the 1PL model for the analyzed dataset.

Graphing the item characteristic curves, item information functions, test information function, and test characteristic curve proceeds then at the software level as in the complete data case. Similarly, the estimation (prediction) of individual trait or ability scores is carried out with Stata using the same commands as in the earlier discussed case of complete data (see chapter 6).

7.7   Chapter conclusion

In this chapter, we were concerned with parameter estimation in IRT models. We discussed at considerable length the role that the popular ML method plays throughout this involved process. In addition, we stressed the fact that in IRT and IRM, we are interested in estimating what may be seen as two "types" of parameters—item parameters and individual subject trait, construct, or ability level (latent score) parameters. We similarly emphasized that each such individual trait level estimate (prediction) is associated with its own SE. Moreover, we also addressed the complicated issue of missing data, specifically demonstrating the utility and applicability of the FIML approach (ML) under the MAR assumption. Finally, we point out also that several additional empirical illustrations of most of the discussion in this chapter were provided in chapter 6.

8 Information functions and test characteristic curves

As was discussed in chapter 6, the concept of information is closely related to the indices of sampling instability of parameter estimates obtained with item response theory (IRT) models, namely, the parameter standard errors. We also indicated earlier that until chapter 11, we will be dealing with binary or binary scored items, that is, dichotomous items, which are widely used for the purpose of evaluating latent constructs, such as abilities or traits of substantive interest in behavioral, educational, and social research. For this dichotomous item setting, we discuss in the present chapter the important concepts of the item information function, test information function, and test characteristic curve (TCC). They will all be instrumentally used in the following chapter, which deals with a multi-item test, scale, or instrument construction and development.

8.1 Item information functions for binary items

When using measuring instruments or item sets consisting of binary or binary scored items, based on the information function related discussion in chapter 6, one can show that the item information function (IIF) for, say, the jth of them, denoted $I_j(\theta)$, is as follows (for example, Baker and Kim [2004]),

$$I_j(\theta) = \left\{P'_j(\theta)\right\}^2 / \left\{P_j(\theta)Q_j(\theta)\right\} \tag{8.1}$$

$(j = 1, \ldots, k)$. In (8.1), $P(\cdot)$ and $Q(\cdot)$ denote the probability of "correct" and "incorrect" response on the item, respectively, as a function of the underlying trait, construct, or ability being evaluated. In addition, priming is used to denote derivative, as in the rest of this chapter. The right-hand side of (8.1) is interpreted as the information provided by item j with respect to the particular level θ (or value of θ) on the latent dimension that the item taps into.

We observe from (8.1) that with a plausible IRT model the information function value can be worked out for any item and ability level θ, once that level or value is specified (fixed, or given). Hence, one can use (8.1) for estimation of the information available in an item with respect to any estimated individual trait or ability level, for example, a particular $\widehat{\theta}$. Further, we stress that in general, information depends on i) the specific ability level θ of concern, because the right-hand side of (8.1) is a function of θ; ii) the item in question, because the ICC determining $P(\cdot)$ in general depends on the item as emphasized by the subindex j in (8.1) (see also next); and iii) the (logistic) IRT model used, because it determines the form of the critical probability $P(\cdot)$ [and thus also the form of $Q(\cdot)$] as a function of θ, which features in the right-hand side of (8.1). ⁐

For the most general logistic model considered in this book for binary or binary scored items, the three-parameter logistic (3PL) model, the right-hand side of (8.1), can be shown with derivative substitution and algebraic rearrangements to lead to the following expression (see, for example, Lord [1980]; compare Birnbaum [1968]),

$$I_j(\theta) = a_j^2(1 - g_j) / \left([g_j + \exp\{a_j(\theta - b_j)\}] [1 + \exp\{-a_j(\theta - b_j)\}]^2 \right) \qquad (8.2)$$

$(j = 1, \ldots, k)$. From (8.2), one observes the following important features of the IIF (fixing all other quantities of relevance that are not explicitly mentioned in each of the five points next):

i) The IIF is higher when the g parameter is small because the expression in the right-hand side of (8.2) is decreasing in the pseudoguessing parameter g.

ii) The IIF is higher when the g parameter is 0, relative to when it is positive.

iii) A positive pseudoguessing parameter g in effect depletes an item of information with regard to any ability level of interest to evaluate; that is, regardless of how high or low a particular θ is, as long as there is guessing, the information in the item with respect to that θ is lower than it would be if there was no guessing on the item.

iv) The IIF is lower for ability levels far from the difficulty parameter b; conversely, the IIF is higher for ability levels closer to the item difficulty parameter (for a given item).

v) The IIF is an increasing function in the item discrimination parameter a, specifically in the vicinity of the item difficulty parameter b; that is, the IIF grows with a. Hence, for a given ability level in that vicinity, items with higher discrimination parameters contain more information about a particular θ than items with lower discrimination parameters.

The discussion in the remainder of this chapter sheds further light on these and related properties of the item information function.

8.2   Why should one be interested in item information, and where is it maximal?

As can be surmised from the preceding developments in the chapter, the special interest in IIF derives from the fact that it represents the contribution that the pertinent item makes to ability estimation at given points along the studied ability continuum. This contribution or information is extremely valuable when it comes to selecting items, from a potentially large pool, for particular measurement or testing purposes. This topic is discussed in more detail in the next chapter.

Our discussion in section 8.1, and in particular (8.2) as well as point v) in it, shows that this contribution depends a great deal on the item's discriminating power, that is, on its a parameter. Specifically, the higher the item discrimination (that is, the a parameter), in principle, the more information there is in the item, particularly, with respect to θ values close to the difficulty parameter, other things being the same (see point v) in the preceding section and below in this one). While this may be intuitively clear in general terms, we need to acknowledge and keep in mind the complex nature of the right-hand side of (8.2) expressing the IIF as a nonlinear function of several quantities that are in a sense "entangled" in it.

As seen from the earlier made point iv) in section 8.1, we emphasize that the item contribution to the information about a trait or ability level, θ, depends also on the item's difficulty parameter. More concretely, the closer the location of interest, θ, to the item's difficulty parameter, b, the higher the information contained in the item about the former (if $g = 0$ holds for the pseudoguessing parameter). Specifically, Birnbaum (1968) showed that an item, say, the jth, provides its maximum information at the following point or location on the studied ability dimension (θ-scale),

$$\theta_{\max,j,\text{guess}} = b_j + \ln\left\{.5 + .5\sqrt{(1 + 8g_j)}\right\}/a_j \tag{8.3}$$

in the general case of a 3PL model being considered ($1 \le j \le k$). Hence, from (8.3), we see that when there is no guessing on the item, the maximum of the information contained in the item is in fact achieved at the point

$$\theta_{\max,j} = b_j \tag{8.4}$$

($1 \le j \le k$). That is, if guessing is not contributing at all to the possibility of coming up with the "correct" answer on an item, then the highest amount of information contained in the item is with respect to the point (location) on the underlying continuum, θ, which equals the item's difficulty parameter, b.

Similarly, we observe from the right-hand side of (8.3) that if there is the possibility of guessing on the jth item, that is, $g_j > 0$, then this item provides the highest information not at the location of its difficulty parameter on the θ dimension (scale) but with respect to a position on the latter that is above that parameter $b_j (1 \le j \le k)$. Thereby, the amount of "positive shift" in this sense of maximal provided information depends on the pseudoguessing parameter associated with the item (for example, Lord [1980]).

8.3   What else is relevant for item information?

In the preceding discussion in the chapter, as well as in the sections of chapter 7 where we dealt with the concept of information, the implicit assumption was made that the used IRT model was correct. In empirical terms, our considerations of item information therefore make an important assumption. Accordingly, the available data in a study of concern are fit well by the item characteristic curves for all items in a measuring instrument of concern, with the curves resulting from a chosen model, or an appropriate "hybrid" model (for example, chapter 11). In other words, the interpretations offered in the above section 8.1 and section 8.2 are contingent upon the selected IRT model being plausible for the analyzed dataset.

At the same time, we need to emphasize also the following point. Even when the fit of an IRT model is good to the study data, an item may still have limited value, if any at all, in all or most tests or multi-item instruments in which it is included. This can happen if the item's a parameter (discrimination power) is low and its pseudoguessing parameter is high. Indeed, revisiting (8.2), we see that when the discrimination parameter a is small while the pseudoguessing parameter g is close to 1, then the right-hand side of (8.2), that is, its IIF value, can be very small. This will be in particular the case when one is interested in evaluating trait or ability levels that are markedly away from the item difficulty parameter b.

This discussion also lets us realize based on (8.2) and (8.3) that the usefulness of an item, in terms of amount of information provided about individual trait, construct, or ability levels, depends actually on the needs that a scale or test developer has when working on or toward constructing a measuring instrument. Specifically, even an item with no or limited guessing and a high discrimination parameter can still be associated with very low information if one were to be interested in ability levels sufficiently far from its difficulty parameter. This is seen from (8.2), as well as (8.3), because the expression there is a decreasing function in the difference $|\theta - b|$, other things being fixed, that is, discrimination and pseudoguessing parameters being constant ($|.|$ is used here to denote absolute value).

Therefore, an item may provide a considerable amount of information with respect to one region of the trait or ability continuum being studied, and at the same time, this item may be next to useless if information is needed at a point or region elsewhere on that continuum. This shows further that item information depends also in a sense on the needs of an instrument (test) developer. More concretely, the amount of information provided by an item depends on the location of the region on the θ scale where a researcher wants to evaluate as well as possible individual trait or ability levels. This fact will turn out to be of particular relevance in the next chapter, where we will be concerned with optimal measuring instrument construction and development. In this connection, we also point out that the use of the term "best item" needs to be well informed. The reason is that as we saw above, an item can be best only when i) the maximum of its information is matched with a region of interest on the underlying ability dimension and ii) the item outperforms there, in terms of this information, other items available to a researcher.

8.4 Empirical illustration of item information functions

The discussion so far in this chapter provided the formal basis for understanding the concept of item information (item information function). As a follow up, we can readily make use of Stata to graphically illustrate this highly important notion, particularly for instrument construction and development purposes (see also chapter 9).

To achieve these aims, we return to the LSAT dataset that we used in chapter 6. As may be recalled, we selected there the one-parameter logistic (1PL) model over the two-parameter logistic (2PL) model as well as over a 3PL model that was fit in that chapter. For the 1PL model, it would thus be of interest to graph the IIFs for the five items analyzed. To this end, upon accessing the data from the file `lsat.dta` (see section 6.1 in chapter 6), we first `quietly` fit again the 1PL model (for its associated output, see section 6.3 there):

```
. use http://www.stata-press.com/data/cirtms/lsat.dta
. quietly irt 1pl item1-item5
```

Once this model is fit, we can use the IIF plotting capabilities of Stata with the following command:

```
. irtgraph iif
```

This yields the graph in figure 8.1 of the IIFs of the five items under consideration.

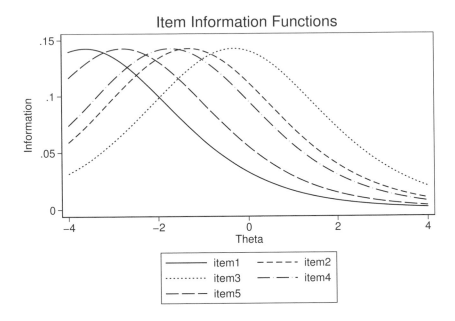

Figure 8.1. Item information functions for the 5 items used (LSAT dataset; one-parameter logistic model)

We observe from figure 8.1 that the IIFs have the same shape and actually seem to have been merely shifted along the horizontal axis or θ scale, by amounts related to the differences in their difficulty parameters (for their estimates, see section 6.3 in chapter 6). This should actually be expected because these IIFs are associated with the 1PL model (Rasch model) fit to the analyzed data, whose characteristic feature is the identity of the item discrimination parameters [see (8.3)]. Further, we easily notice that the information function for item 3 is "positioned" to the right of the other 4 IIFs. This is because its (estimated) difficulty parameter is the largest from those of the five items analyzed.

In concluding this section, we note that if a 2PL model were preferred to the 1PL model in a given empirical setting, then the differences between the IIFs would be more complex. This is due to the discrepancies in the item discrimination parameters. Indeed, as we can see from (8.2), the IIFs are influenced also by the item discrimination parameters that similarly feature in the right-hand side of that equation. Lastly, when guessing contributes to responses on some items, the IIF differences become even more complex, because in that case also, item discrepancies in the pseudoguessing parameters contribute to the former differences.

8.5   Test information function

In sections 8.1 and 8.2, we were concerned with information about a given trait or ability level, θ, that is contained in a single item. However, a behavioral, educational, or social measuring instrument—that is, a scale, test, inventory, self-report, questionnaire, survey, subscale, testlet, etc.—typically consists of multiple items that tap presumably into an underlying trait, construct, or ability of concern. (See chapter 12 for the case of multiple traits or abilities being evaluated by a set of items or an instrument.) Therefore, in such a measurement setting, it is rather natural to ask the following theoretically and empirically important question:

> How much information about a particular location on the underlying latent trait, ability, proficiency, or simply θ-dimension, is contained in a given multicomponent instrument?

This question was first addressed in its entirety in the context of IRT by Birnbaum (1968) on the basis of the general maximum likelihood (ML) theory (for example, chapter 6 and references therein). Accordingly, the total amount of information about a particular trait, construct, or ability level θ, which is contained in the instrument (consisting of k unidimensional items), is given by the following expression:

$$I(\theta) = I_1(\theta) + I_2(\theta) + \cdots + I_k(\theta) \tag{8.5}$$

That is, this total information equals the sum across all items of their information about the location θ of interest on the studied latent dimension. The total information, as defined in (8.5), is referred to as the test information function (TIF). The TIF will be used instrumentally later in this and following chapters. We stress that the TIF is a function of the underlying trait or ability θ, as explicitly stated in (8.5). That is, the amount of information about a particular location on the underlying θ-scale, which is contained in a given set of items or instrument of concern, is not constant but changes in general as that location "moves" along the scale.

An interesting point worth drawing attention to is readily clarified by examining (8.5). Specifically, for a given θ, the test information is accumulated across the items in the instrument or test in question, with each item contributing its "share" independently from any other item. Consequently, the contribution of each item is unrelated to which other items are in the test or instrument (see also below). This is a very useful feature, which results from the general ML theory (for example, Casella and Berger [2002]) and the local independence condition (assumption). (As mentioned earlier, this assumption is advanced throughout the book, as is also usual in most theoretical discussions and empirical applications currently of IRT.)

This discussion leads us to an important related fact that is worth pointing out here as well. In the context of the developments in chapter 7, the estimated standard error (SE) associated with an ability estimate, $\widehat{\theta}$, is here

$$\mathrm{SE} = \mathrm{SE}\left(\widehat{\theta}\right) = 1/\sqrt{\left\{I\left(\widehat{\theta}\right)\right\}} \tag{8.6}$$

$$= 1/\sqrt{\left\{I_1\left(\widehat{\theta}\right) + I_2\left(\widehat{\theta}\right) + \cdots + I_k\left(\widehat{\theta}\right)\right\}} \tag{8.7}$$

This (estimated) standard error is also called the standard error of estimation (referred to for simplicity as SE). We emphasize that it is based on the entire instrument rather than only on a single item. The reason is that the ML theory applies just as well for the estimate $\widehat{\theta}$ obtained using the data from the multi-item instrument (see chapter 6).

In this connection, from (8.7), we see that the standard error of estimation inversely reflects the precision of estimation of (at) the location θ of interest on a studied latent dimension or continuum. [The meaning of the term "inversely" in the last sentence is as stated in (8.7).] We also observe from the right-hand side of (8.7) that this precision i) increases with information about the location θ of concern [whose estimate features critically in (8.7)]; and ii) decreases with the width of the confidence interval associated with $\widehat{\theta}$, which is obtained in the same way as in chapter 7 but using here the quantity SE from (8.7). [See (7.9) in chapter 7 for the general formula that is directly applicable here in tandem with (8.7).]

In simple terms, therefore, the following relationship holds more generally, which is also of fundamental relevance for an empirical application of IRT,

Information $\sim$ Precision of estimation $\sim$ Inverse(SE)

$\sim$ Inverse(confidence interval width)

where the symbol "$\sim$" is used to denote proportionality, that is, an increase of the quantity on the left being associated with an increase in the quantity on the right.

Equation (8.7) reveals another instructive point worth mentioning as well. Specifically, from its right-hand side, it follows that the more information individual items contribute to estimation of a particular location θ of interest on the underlying latent dimension, the smaller the standard error of the resulting ability estimate, $\widehat{\theta}$, and thus the greater the precision of estimation (associated with that estimate). We additionally note that the standard error in (8.7) is not "measurement error" in any physical sense (and not the same as the measurement error commonly conceptualized within the framework of classical test theory—for example, chapter 3). Rather, it is estimation error. Hence, this standard error should not be strictly speaking referred to as measurement error, as has been occasionally done in some IRT-related discussions (see also footnote 1 to chapter 7).

We emphasize in this respect that the standard error in (8.7) results from the general ML theory and a model fit to data, assuming the model is tenable for the latter (see

also chapter 6). We also note from (8.7) that the estimated SE does depend on the
ability estimate. This property is characteristic in general of any ML estimate, not just
individual ability or trait estimates (predictions) obtained within the IRT framework.
In particular, this same property is valid also for any individual factor score estimate
obtained within the ML framework, including those using classical test theory-based
models, classical factor analysis models, or nonlinear factor analysis models, whose
standard errors similarly depend on the underlying trait or ability (see, for example,
Johnson and Wichern [2007]; see also Raykov and Marcoulides [2016b]).

Because the standard-error notion is so important in IRT and item response modeling
(IRM), we emphasize the following features that are also observed from the preceding
discussion in this chapter and in particular (8.7):

i) The larger k, that is, the number of items (assumed homogeneous), and the higher
the information they provide, the smaller the standard error.

ii) The better the items (in general, the higher their a parameters), the smaller the
standard error [see also next point iii)], particularly in suitably chosen ranges of
the underlying θ scale.

iii) The match between item difficulty and a given subject's trait or ability level θ
is also consequential: smaller standard errors come with measuring instruments
composed of items with difficulty parameters close to the ability levels θ that are
of interest to evaluate.

In many empirical situations, it has been found that the size of the standard error
quickly stabilizes as the number of homogeneous items, k, increases beyond a dozen
or so; similarly, the normality of the ML estimates of trait or ability levels is well
approximated with k larger than 20 and in some cases even with $k > 10$ (for example, Hambleton, Swaminathan, and Rogers [1991]). At the same time, it has also been
observed that beyond a certain number of items (say, around 25–30), there are oftentimes diminishing if not almost vanishing returns in terms of lower standard error.
This would be obviously at the expense of including increasingly more items into the
test and enhancing the burden upon the examined person. The last fact is likely to
lead also to compromising the instrument's psychometric quality in an empirical setting (for example, Hambleton and Swaminathan [1985]). We also want to keep in mind
that with increasing length, a measuring instrument may not be functioning in the
way it may be (theoretically) expected to perform. Specifically, i) its validity may in
fact be compromised because of fatigue, boredom, lack of motivation, practicing effects, etc., which are all more likely the longer the overall instrument is (for example,
Raykov and Marcoulides [2011]); ii) the reliability of the overall scale score may also
be diminished by these effects; and last but not least, iii) the instrument may not be
unidimensional any longer (if it was at a shorter length in the first instance).

To illustrate empirically the preceding discussion in this section, we return to the
earlier example with the LSAT dataset (see chapter 6). Given our selection in chapter 6

of the 1PL model for its five items (relative to the alternatively considered 2PL and 3PL models fit there), it is now of interest to look also at the test information function (curve) in (8.5). We achieve this with the following Stata command:

```
. irtgraph tif
```

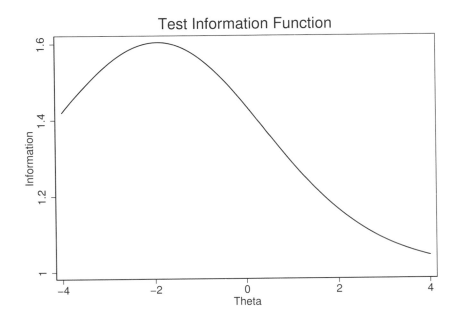

Figure 8.2. Test information function for the set of 5 items used (LSAT dataset; one-parameter logistic model)

This curve shows that the instrument consisting of the five items in question provides most information about subjects with trait or ability levels in the lower end of the presumed underlying general mental ability (GMA) continuum and specifically around two standard deviations below the mean in the studied population. (As mentioned in chapter 6, for the sake of using the example LSAT data for our aims, we assumed it was measuring GMA; we stress that for the present illustration, it is immaterial which particular construct is being evaluated by this five-item instrument.) However, substantially less information is rendered for those persons that are 1 or more standard deviations above the mean on GMA. This observation shows that the instrument would not be nearly as good for that part of the population of concern. That is, by way of comparison, the test consisting of the five LSAT items provides relatively limited information about the GMA levels of persons with notably above average ability.

To see explicitly the magnitude of standard error associated with various positions on the underlying trait continuum, we can superimpose the curve of the former as a function of the latter upon the TIF in this way:

```
. irtgraph tif, se
```

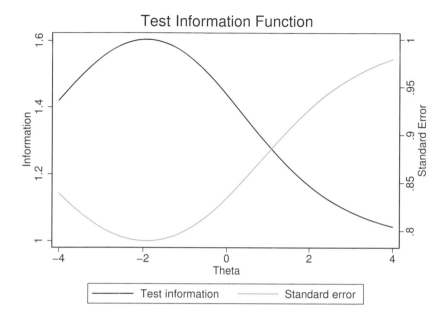

Figure 8.3. Test information function and standard error for the set of 5 items used (LSAT dataset; one-parameter logistic model)

In figure 8.3, we see very clearly the inverse relationship between test information and standard error, which was described earlier in the chapter. (Notice the different scales on the vertical axes for i) the test information on the left and ii) the standard error on the right.) As one approaches on the trait continuum the location of highest TIF values, the standard error becomes minimal. Conversely, the latter increases as one moves away from that location. As was mentioned earlier (see also chapter 7), this behavior is exactly the inverse of that of the TIF, as a function of the underlying trait or ability θ. [For the specifics of this inverse relationship, see (8.7).][1]

1. As an informal example, from figure 8.3, we see that at approximately 1 standard deviation above the mean on the θ-scale, the value of the TIF is (approximately) 1.25, which is associated also with the crossing point of the TIF and the standard-error curve. Hence, following (8.7), the estimated standard error for the underlying ability level at that position on the latent continuum is approximately $1/\sqrt{(1.25)} = 0.89$. This reading (0.89) can be also easily located on the right vertical axis (for standard error) in that figure, at the height of that crossing point of the TIF and standard-error curves. [Observe that the left vertical axis on figure 8.3, for the TIF, is not of direct relevance for the standard-error determination—unless of course the corresponding reading on that axis is square-rooted first and the inverse of the result is then taken; see (8.7). We mention also that the referred being a crossing point of the two curves indicated is not of relevance in this example, and is chosen only for graphical ease in locating it on the figure.]

8.6   Test characteristic curve

As was discussed earlier in the book (for example, chapter 1), the item characteristic curves (ICCs) that are of major relevance in IRT tell us how the individual items function in relation to the underlying trait, construct, or ability that is of interest to evaluate when using a given (unidimensional) set of items or measuring instrument. However, the ICCs are concerned by definition with individual items. Also, as noticed from the earlier developments in this chapter, the TIF is expressed in terms of what may be seen at times as an essentially abstract metric (see, for example, the units on the vertical axis of figure 8.2). In contrast to the ICCs and TIF, the so-called TCC provides this type of information with respect to the entire instrument, test, scale, subscale, testlet, etc., in the original metric of observed scores. Specifically, the TCC tells us overall how the measuring instrument functions in relation to the latent trait of concern in the metric of number of correct (NC) responses.

More formally, the TCC is defined as the expected overall observed score (sum score) at each given value of the underlying trait. That is, the equation

$$\text{TCC} = \mathcal{E}(Y_1 + Y_2 + \cdots + Y_k) \tag{8.8}$$

can be used to define this curve, where $\mathcal{E}(\cdot)$ denotes expectation and the Y's are the observed item responses. In other words, the TCC is the expected NC (or number right) score in relation to the trait or ability presumably tapped into by the item set in question. The TCC evidently differs from the NC in that the latter is always a whole positive number, while the TCC is typically a fractional number because it is the expectation of NC as defined in (8.8).

We stress that just like an ICC, the TCC is a function of the underlying trait, construct, or ability θ. This is perhaps easiest seen by realizing from (8.8) that

$$\text{TCC} = \text{ICC}(1) + \text{ICC}(2) + \cdots + \text{ICC}(k) \tag{8.9}$$

holds, where $\text{ICC}(j)$ is the ICC of the jth item ($j = 1, \ldots, k$). Equation (8.9) follows obviously from the definition of the TCC as sum of the k expectations in the right-hand side of (8.8), while these expectations represent the ICCs of the corresponding individual binary items (see also chapters 1 and 3). Hence, their sum, that is, the TCC, is also a function of the underlying trait or ability, θ, as the ICCs are.

To obtain the TCC associated with a given set of items, to which we have fit a plausible IRT model, we use the Stata command `irtgraph tcc`. To illustrate, let us return to the LSAT example. As will be recalled, for these data, we found in chapter 6 that the 1PL model was preferable to the other logistic models considered there. Once refitting the 1PL model to the LSAT data here, we request the TCC for its five items in this way:

```
. irtgraph tcc
```

The graph produced thereby is presented in figure 8.4.

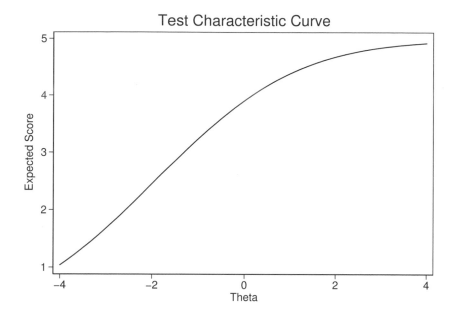

Figure 8.4. Test characteristic curve for the set of 5 items used

We notice from figure 8.4 that the TCC is an increasing, nonlinear function of the underlying trait or ability, θ. This could as well be anticipated because subjects higher on the latent dimension will be expected to provide more correct responses on the items than persons with lower ability. We stress the nonlinear shape of the function in general. This is due to the nonlinearity of the ICCs for the individual items that the TCC is in effect made of [see (8.9)].

Because the TCC was defined as the expected NC score [see (8.8)], in some empirical settings, it may be of substantive interest to a researcher to find out what particular values on the underlying latent dimensions may be associated with prespecified expected NC scores or a range of such. For instance, in the LSAT dataset, it is readily observed that the NC score can range between 0 and 5 because this set stems from 5 binary scored questions. Suppose now that one were interested in the following question: "In which latent continuum range do the latent scores lie for persons who are expected to answer correctly at least 2 but no more than 4 of the items?" To answer it, we use the following Stata command after fitting the preferred 1PL model:

```
. irtgraph tcc, scorelines(2 4)
```

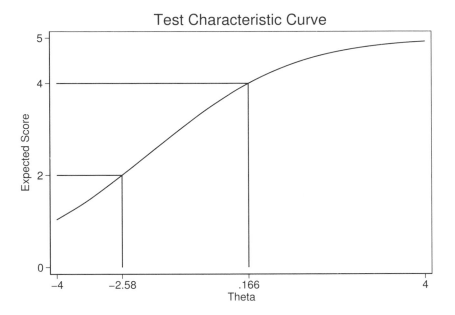

Figure 8.5. Test characteristic curve for the set of 5 items used with latent trait region for expected 2, 3, or 4 correct responses

We see from figure 8.5 that the range of ability levels furnishing expected 2, 3, or 4 correct responses (somewhat loosely speaking) is from -2.58 through 0.166. (Recall that the metric on the underlying latent scale is fixed by setting its mean to 0 and variance at 1.) To be more precise here, given the 1PL model, a latent ability level between approximately 2.5 standard deviations below the mean through 1/6th of a standard deviation above the mean suffices for obtaining an expected NC score between 2 and 4 on the set of 5 items of concern.

Relatedly, a particularly useful, pragmatic feature of the TCC is the following. The TCC also tells what kind of observed sum scores, or NC scores, on the used measuring instrument one could expect from persons with different levels of the studied latent trait (assuming its normality in the population under investigation, an assumption usually made in empirical IRT applications). We obtain that information in this way (see figure 8.6 for the resulting graph):

```
. irtgraph tcc, thetalines(-1.96 0 1.96)
```

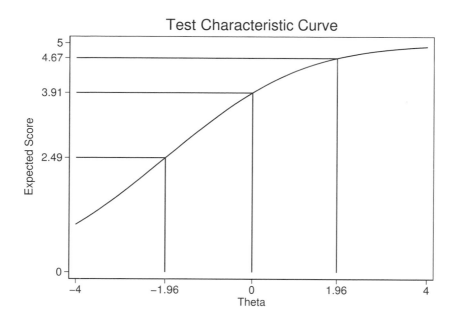

Figure 8.6. Test characteristic curve for the set of 5 items used, with expected score region

The last graph suggests that one can expect above-average individuals with respect to the presumably measured GMA ability to score 3.9 or above on this 5-item instrument. We easily realize, however, that no person could actually score 3.9 on this test; therefore, a more realistic statement would be that we expect above-average individuals with respect to GMA to score 4 or 5 in terms of NC. Similarly, below-average individuals can be expected to score no more than 3.9, that is, realistically no higher than 3 on the instrument (as NC). Last but not least, the TCC graph suggests that 1 person out of 20, in the long run, would be expected not to score between 2.5 and 4.68, that is, not to score 3 or 4. In other words, in the long run, only 1 out of 20 subjects in the population under investigation could be expected to score 0, 1, 2, or 5 as NC on the discussed set of 5 LSAT items.

8.7   The test characteristic curve as a nonlinear trait or ability score transformation

As we observed in chapter 3, with binary or binary scored items, the studied latent trait or ability θ is nonlinearly related to the true score on a given item. Thereby, their relationship is strictly increasing [see (3.8) there]. Earlier in this chapter, the TCC was defined in this setting as the expected NC score on a given set of items or measuring instrument [see (8.8)]. Revisiting now the discussion in section 3.2 of chapter 3, based on (3.8), an alternative reexpression of (8.9) for the TCC is thus the following (see also chapter 1),

$$T_{\mathrm{NC}} = \mathrm{TCC} = \mathrm{ICC}(1) + \mathrm{ICC}(2) + \cdots + \mathrm{ICC}(k) = P_1(\theta) + P_2(\theta) + \cdots + P_k(\theta) \quad (8.10)$$

where T_{NC} denotes the true score for the overall sum score, that is, for the NC score, on the administered set of items. (The discussion in this section assumes that the used IRT model is correct in the studied subject population.)

Equation (8.10) reveals that this strictly increasing, nonlinear relationship between the true score associated with the NC score on the one hand, and the underlying trait or ability that the instrument (presumably) evaluates, θ, on the other hand, can actually be used as a device for scoring individual persons in an IRT modeling session. In particular, upon estimating the individual trait or ability levels $\widehat{\theta}$ (see chapters 6 and 7) and substituting them in the right-hand side of (8.10), we obtain a transformed score for each examined person, denoted $T_{\mathrm{NC}}(\widehat{\theta})$. This score, $T_{\mathrm{NC}}(\widehat{\theta})$, is on the same scale (range from 0 to k) as the NC but, unlike the latter, can also take noninteger values.

To illustrate, returning to the LSAT dataset, let us suppose one was interested in obtaining these transformed scores $T_{\mathrm{NC}}(\widehat{\theta})$ for subjects with identifiers 1, 10, and 20. These persons can be readily found to have their earlier estimated trait or ability level estimates (predictions), $\widehat{\theta}$, correspondingly as -1.91, -1.43, and -0.94 (rounded off). (To remind ourselves, we find these estimates using the Stata command `list Theta in 1/20` after fitting the pertinent 1PL model and requesting the estimation of the individual trait or ability scores, as discussed in detail in section 6.6.) We then use the following Stata command (result presented in figure 8.7):

. irtgraph tcc, thetalines(-1.91 -1.43 -.94)

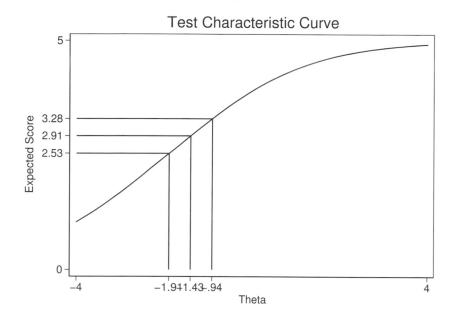

Figure 8.7. Individual transformed scores $T_{\mathrm{NC}}(\widehat{\theta})$ for the three considered persons in the LSAT dataset

We see from figure 8.7 that the transformed scores $T_{\mathrm{NC}}(\widehat{\theta})$ for these 3 persons are correspondingly 2.53, 2.91, and 3.28. We stress that in this straightforward way, we can readily obtain the transformed score $T_{\mathrm{NC}}(\widehat{\theta})$ for any subject whose initial individual trait or ability estimate $\widehat{\theta}$ we know or alternatively find out from the listing of all subject estimates $\widehat{\theta}$ or inquire about in response to a particular research question.

A potential limitation of the transformed score $T_{\mathrm{NC}}(\widehat{\theta})$ is the obvious fact that it is positioned within the interval $(0, k)$. Thus, one may argue $T_{\mathrm{NC}}(\widehat{\theta})$ depends on the number k of items in the instrument or item set used. To alleviate this concern, we can divide $T(\widehat{\theta})$ by the number of items, k, rendering thus the (estimated) true proportion correct score. The latter is sometimes referred to also as domain score (for example, Hambleton, Swaminathan, and Rogers [1991]) and is denoted here $\pi(\widehat{\theta})$:

$$\pi\left(\widehat{\theta}\right) = \left\{P_1\left(\widehat{\theta}\right) + P_2\left(\widehat{\theta}\right) + \cdots + P_k\left(\widehat{\theta}\right)\right\}/k \tag{8.11}$$

Equation (8.11) represents a nonlinear, strictly increasing transformation of the initial individual trait or ability estimate (prediction) $\widehat{\theta}$ for any person in the analyzed sample into an (estimated) true proportion correct score or domain score $\pi(\widehat{\theta})$. Therefore, (8.11) provides another scoring transformation of the initial estimate $\widehat{\theta}$, in addition to i) the T-transformation discussed in chapter 6 that is linear and ii) the above nonlinear transformation $T_{\mathrm{NC}}(\widehat{\theta})$ (see also footnote 3 to chapter 6).

The nonlinear transformation in (8.11) renders an individual transformed score for the underlying trait or ability level, which is always nonnegative and contained in the same numerical interval $[0, 1]$ as the familiar proportion correct score. This feature of the domain score is in contrast to the one associated with the initial estimate $\widehat{\theta}$ that ranges from $-\infty$ to $+\infty$. In this sense, like $T_{\mathrm{NC}}(\widehat{\theta})$, the outcome $\pi(\widehat{\theta})$ of the nonlinear transformation scoring procedure in (8.11) is usually substantially easier to interpret than the estimate $\widehat{\theta}$ in empirical, behavioral, and social research as an individual latent trait or ability score (estimate) resulting from an IRM session.

Thus, when for instance individual pass-fail decisions are to be made, the following approach is readily seen as sensible. While it is difficult to set a cutoff score on the θ-scale, because the product $100 \times \pi(\widehat{\theta})$ ranges between 0 and 100, one may consider using the number 80, say, as a threshold for mastery or proficiency level in pertinent examinations. This approach obviously assumes that the latter number (80, or another relevant one) is justified as such a cutoff based on expert substantive considerations in the subject-matter area of application.

8.8   Chapter conclusion

In this chapter, we capitalized on our earlier discussion of the fundamental concept of information function in IRT. We extended the initially considered item information function concept to that of test information function, explicating them for binary or binary scored items and logistic IRT models. We also discussed the related notion of the TCC, which permits one to interpret overall sum scores in the metric of the popular "number right" (NC) score that is often used and of substantive interest in empirical studies. In addition, we familiarized ourselves with the notion of domain score and related nonlinear transformation, which can be used for individual scoring in an advanced phase of an IRM session. We illustrated discussed concepts on the LSAT example dataset, and they allowed us to also give more pragmatically oriented interpretation of expected NC scores on the set of five items composing it.

9 Instrument construction and development using information functions

On a number of occasions earlier in the book, we indicated a main characteristic of unidimensional item response theory (IRT). Accordingly, this branch of IRT and item response modeling is concerned with a single unobserved continuum along which persons evaluated by a homogeneous set of items are assumed to be positioned. This continuum represents an underlying trait, construct, ability, proficiency, or, more generally, latent dimension that is of special interest to measure (evaluate) using a given set of items that often constitute a multicomponent instrument. Because of this theoretically and empirically highly appealing feature, the result of using (well-fitting) IRT models is an underlying "scale" along which both items and persons are being evaluated (for example, van der Linden [2016b]). Specifically, an application of unidimensional IRT yields person and item parameter estimates (more precisely, item difficulty parameter estimates), which are located on the same scale or continuum that is frequently referred to as the "θ-scale".

It is this feature of IRT that becomes highly useful when it comes to developing (constructing, revising) multi-item measuring instruments like tests, scales, testlets, subscales, inventories, self-reports, questionnaires, surveys, etc., as discussed in the present chapter. This item response modeling property turns out to be especially helpful for selecting items that are most useful in certain regions of the underlying trait or ability scale. These items are supposed to contribute most, from among the members of a prespecified large and homogeneous item pool, to high precision of estimation of item and especially person parameters. This selection process is rooted in properly using the concept of information that the items contribute to the overall instrument information (test information function). That information is critically relevant for the multicomponent instrument to meet its declared requirements and aims, which are generally referred to as specifications, in terms of precision of estimation or "measurement" of the examined persons (see discussion in chapter 8).

As elaborated in the preceding chapter, item and total instrument information are directly related to precision of estimation. Thus, it becomes possible within the (unidimensional) IRT framework to choose items that together produce an instrument possessing the desired precision of evaluation within any given trait, construct, or ability area or range on the θ-scale. How to do this optimally is the main concern of the present chapter.

9.1 A general approach of item response theory application for multi-item measuring instrument construction

The process of optimally constructing multi-item instruments to meet certain measurement quality-related requirements was outlined by Lord (1977, 1980). He described a procedure for using item information functions (IIFs) (see chapter 8) to construct tests having desired specifications, in particular precision of measurement or ability evaluation in certain trait or ability ranges. This procedure is based on the availability of an initial fairly large item pool, often referred to as an item bank (or repository). This pool consists of items known from prior research on the same population of concern to fit well a particular IRT model. The item bank is expected to represent a sufficiently large pool of items of high quality parameters and IIFs that are typically assumed known or previously "calibrated" (evaluated).

Lord's (1977, 1980) procedure is widely used in contemporary behavioral and educational research and especially for (achievement) test construction and development. The procedure consists of the following four main steps (for example, Hambleton, Swaminathan, and Rogers [1991]).

1. Decide on the shape of the desired test information function (see chapter 8). This function, which is based on all available prior knowledge in the substantive domain and population of interest, is the result of expert considerations. The function is best developed with the help of a panel of subject-matter experts, with the specific goals in mind that the resulting instrument is supposed to meet (see additional details below). That function is also oftentimes referred to as the target information function (TaIF).

2. Select items from the above mentioned item bank, which possess the following essential property: they are characterized by IIFs that fill up hard-to-fill areas under that TaIF. This step may be helped by some potentially "what-if" or "trial-and-error" activities that effectively probe different selections of candidate items from the item bank.

3. While engaged in step 2, after each item is added to what is then a working version of the desired instrument, the researcher must calculate the (estimated) test information function (TIF) for all items in it at that time. We stress that this is an estimated TIF because that version of the instrument includes items whose properties (IIFs in particular) have been estimated at an earlier point in

time during item calibration. (Hence, the IIFs available from that process need not be really identical to the "true" IIFs that the items would possess if used on an entire population to be studied with the resulting instrument at the time of its own development.)

4. Steps 2 and 3 are then followed by the researcher who is to continue selecting items from the bank (back-tracking if necessary) until the resulting TIF approximates the TaIF to a satisfactory degree. The latter is to be decided as mentioned also based on expert knowledge and may contain some subjective elements. The extent of lack of complete "filling" of the area under the TaIF at the time when the present four-step process is to be considered completed is to be taken into account when interpreting the results in empirical applications of the instrument constructed using this process.

Next we specialize the above discussion in relation to particular aims that the constructed multi-item measuring instrument is expected to meet.

9.2 How to apply Lord's approach to instrument construction in empirical research

In this application-oriented section, we consider several specific settings and point out how the approach described in the last section (Lord 1977, 1980) can be used to construct instruments with particular measurement-quality properties.

To begin, suppose a researcher is interested in constructing a broad-range ability test (compare Hambleton, Swaminathan, and Rogers [1991]). Such an instrument will have the property of differentiating well between persons that are positioned not particularly low or particularly high on the underlying dimension evaluated by the desired test. Instead, this instrument should furnish sufficiently precise scores for persons that are located in the medium range of the underlying θ-scale. Hence, the researcher will be aiming here at producing a test that provides (approximately) equally and sufficiently precise ability estimates over a fairly wide, intermediate range on the trait or ability continuum. The desirable TaIF to match then will have high and effectively constant values across that range on the θ-scale and possibly be considerably lower outside it. That is, the TaIF should be such that the resulting instrument renders relatively precise ability estimates over a broad range in the medium region of the ability scale at the expense perhaps of less precise trait or ability estimates for persons positioned outside that range of concern (see section 8.3 of chapter 8). This empirical setting may be of particular interest to assessment specialists concerned, for instance, with developing a test that aims to enable good differentiation among most students in a given grade. This may be a goal to pursue in an effort to obtain information about the effect of instruction in a course aimed at meeting certain objectives of relevance for the majority of students except those with fairly low or fairly high levels of knowledge (ability) in the area of instruction. To illustrate this setting, in example 1 in the next section, we present a TaIF that would be desirable to follow or approximate well.

As an alternative scenario, consider the case where a researcher is interested in evaluating low achievers in a population of concern. That is, he or she is interested in developing an instrument that will have its best differentiation capabilities in the lower range of the trait, construct, or ability dimension tapped into by the resulting instrument. Such a situation may be of relevance, for instance, when a school district is interested in constructing a test for identifying as early as possible elementary school students with some learning difficulties. Thus, the researcher will need to work with a TaIF that is high and approximately constant only on a relatively limited range in the lower region of the ability scale. In example 2 in the next section, we illustrate such a case of empirical interest.

Yet another setting of relevance oftentimes in achievement-related contexts is found when a researcher is conversely concerned with developing an instrument designed to differentiate well among high achievers. That is, he or she is interested in the optimal differentiation between subjects that possess high levels of the studied trait or ability. This can be the goal, for instance, when developing an instrument to aid university-level decision makers with respect to awarding prestigious fellowships. Then the TaIF of importance would have a correspondingly modified form, namely, being high and approximately constant in the upper region of the θ-scale. We illustrate this scenario in example 3 in the next section.

Last but not least, for a criterion-referenced test, for example, a licensure examination, with a prior established and substantively defensible cutoff score C to separate "masters" from "nonmasters", one would desire very high precision in a relatively limited range of the θ-scale that covers C. That is, in this setting, a researcher would be interested in differentiating well between proficient versus nonproficient applicants for obtaining the license in, say, a professional field of concern. Hence, the TaIF of relevance to match should be highly peaked near and around the position C on that scale. We illustrate this scenario in example 4 in the next section.

This discussion demonstrates a more generally valid principle that is effectively also embedded in the approach described by Lord (1977, 1980) and indicated in section 9.1. Specifically, suppose one had a good idea of the trait or ability level (range) of a group of persons of particular interest. Based on their associated information functions, candidate items can then be selected from an available item bank to maximize the test information in that region of the trait of relevance, which is spanned by the subjects to be measured with the resulting instrument. Moreover, the TIF should also be of i) relatively uniform (constant) and marked height across that ability range of interest and possibly ii) higher (or even notably higher) than its values in the rest of the θ-scale. That is, the concern is in effect exclusively with ensuring high and comparable precision of estimation of the trait or ability levels within the region of relevance. This can then be achieved at the expense of lower precision for persons with ability levels outside that region (see section 8.3 of chapter 8).

We emphasize that the reason the discussed application of the procedure outlined in the preceding section is possible is an earlier mentioned characteristic and highly useful feature of (unidimensional) IRT. Accordingly, items (difficulty parameters), persons, and substantively meaningful cutoff scores where applicable or available are all measured on or associated with the same underlying scale (ability or trait dimension), the θ-scale. We also stress that an essential assumption of this procedure is obviously the availability of trustworthy estimates of item parameters for the elements of the item bank, specifically prior estimates with fairly small standard errors. For this to be the case, a necessary condition is that their earlier calibration or previous estimation must be carried out on a representative and large sample of subjects from the studied or intended population to investigate.

9.3 Examples of target information functions for applications of the outlined procedure for measuring instrument construction

As indicated earlier, in this section, we present graphical examples for TaIFs that correspond to the several settings discussed in the preceding section (compare Lord [1980] and Hambleton, Swaminathan, and Rogers [1991]). The following examples should best be treated as providing suggestions to consider particular TaIF shapes, rather than strict guidelines or prescriptions to be followed rigidly in applications. We also point out that in the figures to follow i) the horizontal axis represents the ability scale (that is, the θ-scale); ii) the vertical axis represents the test information; iii) the origin of the vertical axis may but need not be at the axes' crossing point; and for the sake of general applicability iv) that origin point is kept unspecified on the horizontal axis (θ-scale), unless stated otherwise. With this in mind, the following figures display possible target test information functions for several settings where a researcher would be interested in constructing measuring instruments with different aims. (That is, the instrument specifications differ across the following four settings.)

▷ **Example 1**

When constructing a broad-range ability test, let us suppose one is interested in developing an instrument that produces (approximately) equally highly precise ability estimates across a relatively broad range of ability. Then a TaIF that may well be worth considering would look like that in figure 9.1. In the figure, the range of the ability scale in which the TaIF is highest can be chosen, for instance, to be $(-2, 2)$ or even $(-3, 3)$. The particular choice depends on the researcher's decision with respect to how broad the desirable range of high-precision estimation would need to be.

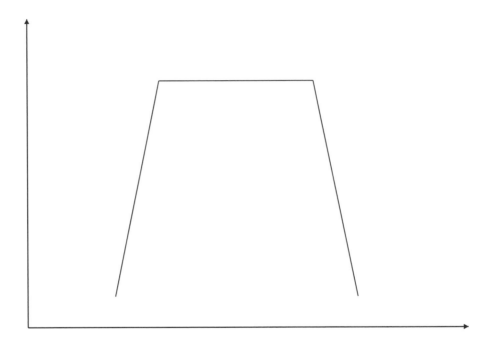

Figure 9.1. Example of a target information function for a broad-range ability test

◁

▷ **Example 2**

When developing a test to evaluate low achievers, let us suppose a researcher is interested in differentiating between them, say, with respect to their achievement in a given elementary school curriculum subject. In that case, a TaIF like the one in figure 9.2 may well be appropriate to consider.

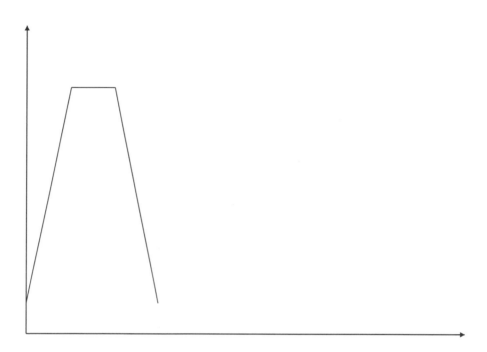

Figure 9.2. Example of a target information function to evaluate low achievers

◁

▷ Example 3

Alternatively, when constructing a test to screen high achievers and differentiate between them, let us suppose a TaIF like the one in figure 9.3 may be considered.

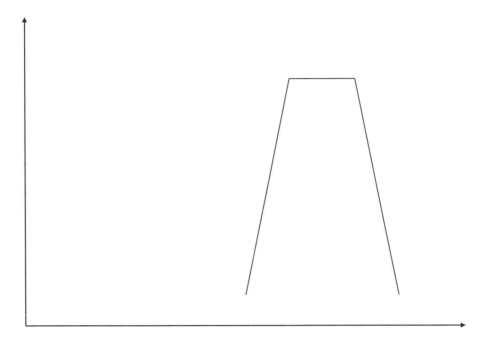

Figure 9.3. Example of a target information function to evaluate high achievers

◁

▷ Example 4

When developing a criterion-referenced test, let us suppose a scholar is given a substantively justified cutoff C on the presumed underlying latent continuum, the θ-scale, which the items used thereby should tap into. (Recall that for model identification purposes, the unit on that scale is the population standard deviation, and the scale origin is the population mean.) Assuming C is at the place in figure 9.4 where the maximum of the TaIF is positioned, for its shape, a researcher may consider one that approximates well the curve displayed on the figure.

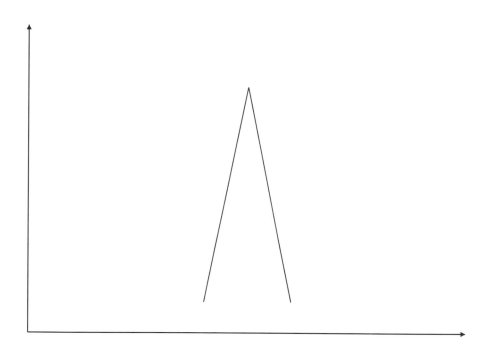

Figure 9.4. Example of a target information function for a criterion-referenced test

◁

We readily observe that all four figures of TaIFs have one important feature in common (see chapter 7). Specifically, the height of the target information function is the inverse of the required (acceptable or desirable) standard error for the range of interest, with somewhat higher standard errors allowed outside that range. In particular, as we know from chapter 7, the information function at any given point on the θ-scale is inversely related to the standard error of estimation (SE) at that point [see (9.1) for more detail]. That is, dropping the argument θ for simplicity, we see that

$$I = 1/\text{SE}^2 \tag{9.1}$$

Suppose now that an instrument developer decides that an SE $= 0.25$, say, is acceptable in the trait or ability range of interest. This will be the case if he or she determines based on substantive considerations that the acceptable width of, say, the 95% confidence interval for the trait level within that range is approximately $1 (= 4 \times 0.25)$. (See chapter 5 regarding the induced "metric" on the θ-scale, in which the unit equals the variance of the underlying trait or ability of interest to evaluate.) Then from (9.1) follows that the maximal height of the TaIF should be (approximately) equal to $I = 1/(1/4)^2 = 16$.

9.4 Assumptions of instrument construction procedure

The instrument construction and development procedure discussed in this chapter rests on several important assumptions that were indicated earlier but need to be specifically emphasized, as we do in this section.

First, it is essential that a sufficiently large and representative sample from the population intended to be studied subsequently was used when the initial item bank was created. This is needed to ensure that the item parameters are estimated with high precision, that is, are associated with small standard errors. In this connection, one needs to keep in mind that even if such a sample was obtained at a given earlier point in time, particularly in an educational setting (as well as others), it would be hard to expect that population not to have changed over time in a way that may be potentially important substantively. Hence, at the time when the instrument needs to be constructed as described above, that sample's data may not be really relevant anymore. Thus, the item parameters that were estimated (calibrated) on the earlier sample as well as (some of) their item information functions may no longer be sufficiently trustworthy at the time of actual instrument construction. This is because one may not be able to assume then that "a large enough item bank with known measurement properties" is indeed available for this purpose.

Second, and relatedly, populations of interest in the behavioral, educational, and social sciences as well as beyond them evolve with time and, in particular, tend to become typically more diverse and heterogeneous. Therefore, large samples from them may include subjects belonging to substantively different "classes", or mixtures, to begin with. This is likely to have at least some effect on the optimality claimed for the instruments obtained as described earlier in this chapter. Under such circumstances, the use of mixture IRT at the time of item calibration is definitely recommended to consider (see, for example, van der Linden [2016b]; see also Raykov, Marcoulides, and Chang [2016]).

Third, the instrument construction procedure discussed in this chapter is based mostly on statistical criteria. Therefore, because of their essentially formal nature, they need not be in a position to ensure content, criterion, or, more generally, construct validity (for example, Messick [1995]). Hence, unless convincing arguments are provided in favor of validity of the resulting multicomponent instrument, it would be in general unknown to what extent it possesses such validity characteristics. In particular, it may not be immediately obvious to what degree the instrument may be affected by "construct underrepresentation" and "construct-irrelevant variance" (for example, Raykov and Marcoulides [2011]). Therefore, it is essential to ensure careful, judicious, expert, and validity-focused item choice from the item bank during instrument construction so that the validity of the produced measuring instrument with the above procedure is not compromised. This requires very careful application of Lord's procedure in empirical settings, paying thereby special attention to validity issues as well as to test length. The latter point may be particularly important to also keep in mind because, in an effort to avoid construct underrepresentation, one may be inclined to construct longer tests that may in turn suffer from validity loss in practical settings. This loss may result from undesirable effects of excessive test length, such as fatigue, boredom, or loss of interest, motivation, or attention by the examined subjects.

Fourth, another important assumption underlying the discussed instrument construction procedure is that the item bank consists of a relatively large pool of high-quality items that are "alike". That is, the items must be "comparable" with each other in the sense of being homogeneous (unidimensional). To be on the safe side, one may argue, however, that longer tests are more likely not to be strictly unidimensional. This may be due to inadvertently emerging "minor" traits or abilities with increasing test length, which are also tapped into by some of the items. Therefore, examination of instrument homogeneity needs to be carried out at least at the end of an application of the above test-construction procedure. It is even preferable to ensure unidimensionality throughout the process of creating the item bank in the first instance. (For a discussion of the issue of instrument multidimensionality, see, for example, chapter 12.)

9.5 Discussion and conclusion

In this chapter, we discussed an information function-based, and hence estimation precision-based, procedure for construction of multicomponent measuring instruments (Lord 1977, 1980). The approach relies on strong requirements, as we have also pointed out (see section 9.4). However, when they are fulfilled or at least plausible in empirical settings, the method yields instruments with optimal precision for the aims they are developed. Thus, instruments constructed using this approach are often called optimal tests. This is also because they are associated, as we saw, with optimal (highest) precision of estimation of ability levels in prespecified ranges.

These optimal measuring instruments usually differ markedly from what may be called "standard instruments". The latter may well be still available in some areas of the behavioral, educational, and social disciplines. In fact, standard instruments may have been in part (or completely) "randomly" constructed in earlier years. Research comparing optimal instruments resulting from the above procedure on the one hand and standard tests on the other has shown that over the range of interest, the precision of the optimal tests could be achieved by standard tests that are considerably longer than corresponding optimal tests (see, for example, Hambleton, Swaminathan, and Rogers

[1991]; see also Lord [1980]). That is, the relative efficiency of the optimal tests is markedly higher in that range of relevance. However, it is possible that standard tests perform as well or even better than optimal tests outside the specific ranges of ability levels of concern and focal interest for the latter instruments (for example, Hambleton and Swaminathan [1985]). Therefore, it is particularly important to be well aware of and explicit about the special ranges for which optimal tests have been or are being developed. This will be essential when ensuring that they are used in empirical research exclusively for measuring traits, constructs, or abilities for persons with individual levels that are covered by those specified ranges for the optimal tests.

10 Differential item functioning

While educational and behavioral assessment may be considered to be more than a century old, an issue in mental measurement and testing that has been possibly of greatest importance to the public in recent decades is that of equity or fairness (for example, Holland and Wainer [1993]). Specifically, publishers of high-stake tests and examinations as well as their official users are expected to demonstrate that their tests are free of bias against particular groups. These groups might include, for example, individuals of different ethnic or cultural origin, gender, socioeconomic status, or other characteristics (subpopulations). Evidence in favor of fairness with respect to these or other distinct groups of concern in a given empirical setting should therefore always be supplied before recommending wider use of an educational or behavioral measuring instrument.

To address this consequential problem of test bias, one must first gather sufficient empirical evidence with respect to the relative performance on individual items and the overall test or instrument. In general terms, this evidence has to be collected by the instrument or test publisher or related institutions from i) individuals of different ethnic or cultural groups, including minority and majority group members; ii) potentially discernable advantaged or disadvantaged group members; or iii) any other possibly relevant subpopulations of interest. We stress, however, that gathering such empirical evidence may be only necessary but not sufficient to arrive at a more general conclusion that bias in a measuring instrument is not present with respect to these groups, or subpopulations (compare Osterlind and Everson [2009]). In particular, a conclusion for lack of such bias, and hence fairness to all groups expected to be evaluated by the test, may involve inferences that go beyond the available data and associated statistical analyses.

10.1 What is differential item functioning?

The typical use of the term differential item functioning (DIF) refers to empirical evidence pointing toward item or overall test bias. This evidence is usually collected in the process of examining the way a measuring instrument and its components function when evaluating different groups (that is, different subpopulations as mentioned above). To give a more formal definition of DIF, we need to introduce first some important terms that will be used throughout the rest of this chapter. Specifically, for convenience, we will refer to the "majority" or (presumably) "advantaged" group or subpopulation from an overall population of interest as a "reference" group. For example, in some settings, this might possibly be male examinees. Conversely, we will refer to the other "minority" or "disadvantaged" group as a "focal" group. For instance, in some settings, this might possibly be female examinees. Also, as with the preceding discussion in this book, we assume in the present chapter that an item of concern is binary or binary scored. We presume thereby that its answers are scored 0 or 1 as before and, for convenience, will refer to the latter as "correct" response. [See chapter 11 for polytomous items and corresponding item response theory (IRT) models].

An informal description aimed only at giving a rough idea of what may be related to DIF with respect to an item in a multicomponent measuring instrument or an item set, could be as follows (compare, for example, Hambleton, Swaminathan, and Rogers [1991]): an item shows DIF if the reference and focal groups differ in their mean performance on the item. While this description provides a first glimpse into DIF, it is not precise and in fact not recommended to be used as a DIF definition. The issue with this DIF description is that it does not say anything about the possibility of a real between-group difference existing on an underlying trait, construct, or ability of actual interest, which is tapped into by an item in an instrument or test of concern. (As indicated earlier in the book, in this chapter, we also assume that this trait, construct, or ability is unidimensional and is as usual denoted θ; see chapter 12 for the alternative case.) Yet it is this latent difference that may in fact be responsible for the observed group difference on a given item or overall measuring instrument, while it is "left out" in the above description of DIF.

Therefore, the following definition of DIF is widely accepted among psychometricians and underlies this chapter (compare Holland and Wainer [1993]). Accordingly, an item shows DIF if individuals with the same trait or ability level, but belonging to different groups (subpopulations), do not have the same probability (mean) of responding "correctly" to the item. That is, if for an item, the inequality

$$P_f(Y = 1|\theta) \neq P_r(Y = 1|\theta) \tag{10.1}$$

is true for at least one value on the latent trait or ability dimension, where the subindexes "f" and "r" indicate correspondingly the focal and reference group and Y denotes the observed score on the item, then the item and hence the measuring instrument including it are associated with DIF. Note that if (10.1) holds, then the same inequality applies for the conditional probability for response "0", given any ability or trait level θ.[1]

Alternatively, if for all θ (that is, for all real values of θ) the two conditional probabilities appearing on either side of (10.1) are the same, that is, that inequality does not hold, then the item of concern is not associated with DIF. (In this case, obviously, the two respective conditional probabilities for response "0" will be identical across the

---

1. Because the underlying ability or trait dimension evaluated by an instrument of concern is (assumed to be) a continuum, and owing to the continuity of either item characteristic curve (ICC) involved and thus of their difference, the following statement is obvious: if (10.1) holds for one value or point θ, there will be infinitely many ability or trait levels (scores, θ values, or points on the θ-scale) for which (10.1) will hold as well, namely, all those in a sufficiently narrow interval around θ, that is, in a sufficiently small neighborhood of θ (for example, Zorich [2016]).

groups in question.) In other words, if it does not make any difference for the probability of "correct" response on the item which group someone belongs to, who has any prespecified ability or trait level θ, then this item is not associated with DIF. For a given measuring instrument, such as a test or scale, a single item with DIF is sufficient to declare the instrument itself exhibiting DIF. The instrument will not exhibit DIF only when there is no item in it that exhibits DIF.

Based on the discussion provided in earlier chapters of the book (for example, chapters 1 and 3), because the two probabilities appearing on either side of (10.1) are actually the ICCs for a considered binary or binary scored item in the focal and reference group, respectively, it follows that the question of DIF is equivalent to asking whether these ICCs are the same in the groups. [Note that one can reduce the comparison among more than two groups to several comparisons of two groups, for which these ICCs become then of relevance in relation to DIF. See below for a multiple testing method, the Benjamini–Hochberg (BH) procedure, which will also be applicable then.]

Inequality (10.1) obviously involves two probabilities that are functions of the underlying trait or ability being tapped into by an item of interest. However, the behavior of these functions, as θ takes all of its possible values, is not restricted (other than the functions yielding probabilities). That is, these functions can have in general any shape in the groups. Thus, when the two group-specific ICCs are not identical, there are two qualitatively distinct ways in which they can differ. Therefore, two possible forms of DIF can occur that are called uniform and nonuniform DIF. Specifically, when the two ICCs cross over, we speak of nonuniform DIF. Conversely, when they do not cross over while (10.1) still holds, we have a case of uniform DIF. We stress that either of these forms is equally relevant as evidence for DIF. That is, either of these two forms being the case for a given item is sufficient for the latter to be exhibiting DIF.

With the preceding discussion in this chapter, we are now ready to move on to a description of statistical approaches for the study of DIF.

10.2 Two main approaches to differential item functioning examination

There are two main approaches to studying DIF, which are related yet formally distinct: the i) observed variable-focused and ii) latent variable-focused methods for DIF examination (compare Holland and Wainer [1993]). In the former, group differences on the observed items are at the center of attention. Usually, the overall sum score (total score, or "number correct" score) is taken as a proxy of the trait, construct, or ability being evaluated. We will refer to these procedures for convenience as "observed variable methods" in this chapter. Alternatively, in the methods mentioned in ii) latent-variable models—such as IRT models—are explicitly used (compare Raykov et al. [Forthcoming]). In these models, the studied trait, construct, or ability is presumed responsible for subjects' performance on the items in a studied instrument or test. This latent variable is explicitly involved in the process of DIF examination. We will refer to these methods as "latent-variable methods" in the rest of the chapter.

In the observed variable methods, logistic regression or related procedures are typically used. In the latent-variable methods, a latent-variable model including latent variables is postulated in either group and comparison of its parameters is essential for studying DIF. The remainder of this chapter will consider procedures for examining DIF that are available with Stata as the software underlying this book. We discuss the two types of methods in turn next and refer to Holland and Wainer (1993) for further details and related procedures for studying DIF.

10.3   Observed variable methods for differential item functioning examination

Within this set of methods, two procedures have become popular. These are logistic regression and the Mantel–Haenszel (MH) test (Agresti 2013). In logistic regression, as indicated earlier, the total score on the (unidimensional) instrument or test, denoted "t" in this section, is used as a proxy for the underlying trait being evaluated. That is, a characteristic feature of this approach is the use of an observed score as assumed to be indicative of the underlying latent trait or ability evaluated by a multi-item measuring instrument of concern.

To make use of logistic regression for examining DIF, we can apply the three models defined in the following: (10.2), (10.3), and (10.4) (compare *Stata Item Response Theory Reference Manual* [2017a]). They are nested in each other and allow a researcher to examine both nonuniform and uniform DIF for a given item. In these equations, the probability of "correct" response on the item is denoted by "p", group membership is symbolized by "g", and the interaction of total score ("number correct" or "number right" score) with group is designated "$t \times g$" (see chapter 2 for the definition of the log odds or logit of a given probability):

$$\text{Model 1: } \text{logit}(p) = \beta_0 + \beta_1 t + \beta_2 g + \beta_3(t \times g) \tag{10.2}$$

$$\text{Model 2: } \text{logit}(p) = \beta_0 + \beta_1 t + \beta_2 g, \text{ and} \tag{10.3}$$

$$\text{Model 3: } \text{logit}(p) = \beta_0 + \beta_1 t \tag{10.4}$$

In (10.2) through (10.4), the β's denote the intercept and partial logistic regression coefficients within the models. (For simplicity and convenience, the same notation for these coefficients is used across equations.) We emphasize that an interaction of total score and group will be relevant to include in a model for DIF (model 1) when the ICCs for the two groups intersect. We also note that model 2 is obtained from model 1 when the null hypothesis $H_{0,1}: \beta_3 = 0$ is true (plausible). Further, model 3 is rendered from model 2 when the null hypothesis $H_{0,2}: \beta_2 = 0$ is true (plausible). Similarly, note that there is no DIF only when model 3 is preferred to both models 1 and 2. That is, there is no DIF on the studied item when both null hypotheses $H_{0,1}$ and $H_{0,2}$ are true (plausible). The reason is that the last term on the right-hand side of (10.2) (model 1) captures nonuniform DIF, while the last term on that side of (10.3) (model 2) picks uniform DIF.

Because the above three binary logistic regression models are nested in each other, we can use the likelihood-ratio test (LRT) to study DIF with this procedure. Specifically, a test of nonuniform DIF is furnished by testing model 1 against model 2. If the latter is preferred, one can argue in favor of no nonuniform DIF. However, in that case, the possibility of uniform DIF still remains. To test for uniform DIF, we compare model 2 with model 3. This is also possible using the LRT. We note that according to the LRT theory, each test statistic used thereby—which equals (-2) times the difference in the maximized log likelihoods under the models involved (see chapters 6 and 7)—is distributed with large samples, under the null hypothesis being tested, following a chi-squared distribution with 1 degree of freedom (Agresti 2013). We also observe that if we reject with the first-mentioned LRT application the null hypothesis of no nonuniform DIF, that is, $H_{0,1}$, there is no point in testing for uniform DIF using the second-mentioned LRT statistic. That is, in this case, it is pointless to test also the hypothesis $H_{0,2}$, because one already has (statistical) evidence for DIF as a result of the preceding LRT application leading to the rejection of the null hypothesis $H_{0,1}$.

Next we illustrate the outlined logistic regression-based procedure on empirical data.

10.4   Using Stata for studying differential item functioning with observed variable methods

The software Stata implements the observed variable method discussed in the last section and is readily used for examining DIF with it (see also *Stata Item Response Theory Reference Manual* [2017a]). To demonstrate this application, we use adapted data from an ability test consisting of $k = 9$ binary scored items. The test was administered to $n_1 = 771$ male and $n_2 = 744$ female high school students, and its data are found in the file dif2.dta (compare De Boeck and Wilson [2004]). We read in these data first, take a look at the variables, and tabulate the gender variable that is denoted as female (where tabulate could be shortened to tab):

```
. use http://www.stata-press.com/data/cirtms/dif2.dta

. describe

Contains data from http://www.stata-press.com/data/cirtms/dif2.dta
  obs:          1,515
 vars:             10                              25 Oct 2016 10:02
 size:         60,600
```

| | storage | display | value | |
variable name	type	format	label	variable label
item1	float	%9.0g		
item2	float	%9.0g		
item3	float	%9.0g		
item4	float	%9.0g		
item5	float	%9.0g		
item6	float	%9.0g		
item7	float	%9.0g		
item8	float	%9.0g		
item9	float	%9.0g		
female	float	%9.0g		

```
Sorted by:

. tabulate female
```

female	Freq.	Percent	Cum.
0	771	50.89	50.89
1	744	49.11	100.00
Total	1,515	100.00	

To examine uniform and nonuniform DIF with the observed variable procedure out-lined in the preceding section, we use Stata's available command `diflogistic`:

```
. diflogistic item1-item9, group(female)
Logistic Regression DIF Analysis
```

| | Nonuniform | | Uniform | |
Item	Chi2	Prob.	Chi2	Prob.
item1	1.35	0.2450	14.17	0.0002
item2	1.48	0.2244	1.59	0.2067
item3	0.42	0.5191	6.25	0.0124
item4	8.07	0.0045	3.84	0.0499
item5	2.56	0.1094	5.68	0.0171
item6	1.17	0.2798	0.26	0.6116
item7	0.01	0.9264	2.90	0.0887
item8	0.87	0.3511	1.89	0.1694
item9	0.32	0.5698	2.28	0.1308

The last presented output with the LRT statistics and associated p-values seems to suggest, to be on the conservative side, that i) item 4 may be associated with nonuniform DIF, while ii) item 1 with uniform DIF (see next for qualification). This is a rather informal suggestion, however, that cannot be considered conclusive because we have in effect conducted multiple testing on the same dataset. Indeed, this output implies that 18 statistical tests have been carried out on the analyzed dataset with respect to the 18 underlying logistic regression parameters of relevance here (9 interaction and 9 group parameters), as discussed in section 10.3.

To account for this multiplicity of testing, we can use the increasingly popular Benjamini–Hochberg (BH) multiple testing procedure that is based on the false discovery rate concept (see Benjamini and Hochberg [1995] and see Raykov, Lichtenberg, and Paulson [2012] for a nontechnical description, or the Appendix to this chapter). When applied on the above 18 p-values associated with these 18 tests, this procedure determines that only 1 of their pertinent null hypotheses should be rejected. (See section 10.5 for details on the application of this procedure, as well as the Appendix.) Specifically, this is the hypothesis associated with the smallest p-value of 0.0002. This null hypothesis stipulates that item 1 is associated with no uniform DIF. Therefore, we can conclude that the analyzed data contains sufficient evidence to reject the hypothesis that item 1 does not exhibit uniform DIF. That is, we can conclude that the overall measuring instrument consisting of the 9 items under consideration exhibits DIF, with item 1 in particular showing uniform DIF between male and female high school students.

It is still unknown, however, which of the two gender groups may have been "favored" by this item. To explore the direction of such "favoring" by item 1, we can use the following command:

```
. tabulate item1 female
```

| | | female | |
	item1	0	1	Total
0	309	274	583
1	462	470	932
Total	771	744	1,515

We can see informally from this table that females have disproportionately more 1s (and hence disproportionally fewer 0s) than males on this item. Hence, females seem to be the "favored" group on the item. To be more precise, because item 1 shows uniform DIF, we can use the popular MH test to evaluate its degree of DIF (for example, Agresti [2013]). This test accumulates across all individual observed total scores the chi-squared test statistic for the contingency table of gender by item at each such score, that is, after conditioning on the number of correct scores. (Typically, as can be implied from its description, the MH test can sense uniform DIF, but this is not relevant here because we already tested for both uniform and nonuniform DIF; see above in this section.) In addition, and of particular importance now, this approach allows one to obtain point and interval estimates of the extent and direction of DIF. These estimates are provided by those of the pertinent common odds ratio and may be especially informative in an empirical setting. We conduct the MH test with Stata using the command difmh (as mentioned, we are concerned only with the result for item 1):

```
. difmh item1-item9, group(female)
Mantel-Haenszel DIF Analysis
  Item |     Chi2      Prob. | Odds Ratio    [95% Conf. Interval]

  item1 |    13.40     0.0003 |    1.6296      1.2594      2.1085
  item2 |     1.55     0.2130 |    1.1673      0.9255      1.4723
  item3 |     5.99     0.0144 |    1.4292      1.0809      1.8895
  item4 |     2.94     0.0863 |    0.8123      0.6443      1.0242
  item5 |     4.87     0.0273 |    0.7063      0.5239      0.9522
  item6 |     0.29     0.5892 |    1.0812      0.8398      1.3919
  item7 |     2.02     0.1557 |    0.8274      0.6444      1.0622
  item8 |     1.76     0.1849 |    0.7937      0.5761      1.0935
  item9 |     2.14     0.1435 |    0.8265      0.6466      1.0565
```

We notice here that the odds ratio associated with item 1 is estimated at 1.63, rounded off (see middle entry in the first row of the last presented output section). Further, this ratio is notably above 1 because its 95% confidence interval is $(1.26, 2.11)$ and thus positioned entirely above 1, as we observe from the last 2 entries in the first row of that section (see also Agresti [2013].) We thus conclude that the analyzed dataset contains evidence suggesting that item 1 is favoring the focal group, that is, females, which is the group with 1 on the variable `female` (for example, *Stata Item Response Theory Reference Manual* [2017a]).

10.5   Item response theory based methods for differential item functioning examination

The observed variable methods discussed in the preceding section do not make reference to or use of a (presumed) latent trait, construct, or ability dimension of interest that is being evaluated with an item set or multi-item measuring instrument of concern. In addition, those methods condition on the total score, which could rarely be considered a sufficiently good or precise measure of the underlying trait of actual relevance and typically contains potentially sizable measurement error. When one wishes to study DIF while being concerned also with the trait itself and thus accounts for that error, as may well be recommended, IRT offers alternative methods not sharing the above limitation (compare Osterlind and Everson [2009]). They are based on the previously mentioned fact that DIF is equivalent to the lack of complete identity across groups of the ICCs associated with each item in the set or instrument considered. Hence, when one uses binary or binary scored items that are homogeneous and with no (or only minimal and negligible) guessing, lack of DIF is equivalent to the group invariance in the discrimination and difficulty parameters in the two-parameter logistic model. The latter is then the most general logistic model that we have used in the book. (The following method is directly applicable to the one-parameter logistic model as well, after corresponding minor adjustment; see below.) Hence, in this setting, we can examine DIF via the following IRT-based approach that consists of several steps (compare Raykov et al. [Forthcoming]):

Step 1: Fit the IRT model, denoted as M0, to the given set of, say, k items $(k > 1)$ of concern, which is to be treated as a baseline model for the remaining analyses and is characterized by the following two features:

a) fixing of the latent mean and variance at 0 and 1, respectively, in one of the groups (for example, the reference group) but leaving them free in the other group (the focal group); and

b) constraining for group identity the corresponding discrimination and difficulty parameters for all items.

Step 2: Releasing a discrimination or threshold parameter in M0, one at a time, work out via use of the LRT the corresponding p-value associated with the null hypothesis that this discrimination or difficulty parameter is the same in the groups. (Do not interpret any of the $2k$ resulting p-values at this stage; see next.)

Step 3: On this set of $2k$ p-values, apply the BH multiple testing procedure (see Appendix and Wasserman [2004]).

Step 4 (conclusion step): If none of the tested $2k$ hypotheses is to be rejected according to the BH procedure conducted in step 3, there is no indication for DIF on any of the items and thus the entire instrument under consideration. However, if at least one of these $2k$ hypotheses is rejected, there is evidence of DIF for the instrument. The source of DIF is then located in those item discrimination or difficulty parameters, which are associated with the rejected hypotheses of group identity.

We illustrate this IRT-based DIF study procedure using an adapted dataset from a psychometric test consisting of $k = 9$ binary items administered to two groups of 761 males and 739 females (compare De Boeck and Wilson [2004]). In the pertinent data file dif1.dta, the items are named q1 through q9. Further, the male group is coded 0 on the variable female there and is used as a reference group in the following analyses. Also, the female group is coded 1 on that variable and is used next as a focal group. To use the above DIF study procedure, we make use of Stata's feature to process batch files, referred to as do-files, after we access the last mentioned Stata data file (see, for example, chapter 6 for how to do it). To enable this feature, first we start the so-called do-file editor with the following command:

```
. doedit
```

In the left window, we enter the following sequence of commands that will fit model M0 from step 1 above in this section (we are grateful to Dr. Raciborski for providing the original version of the following Stata do-file, which was modified to a minor extent to meet the present purposes):

```
use http://www.stata-press.com/data/cirtms/dif1.dta
* create group-specific item labels with the following 4 lines
forvalues i = 1/9 {
  generate byte q`i´_f = q`i´ if  female
  generate byte q`i´_m = q`i´ if !female
}
* Model M0: For each item, constrain its difficulty and
* discrimination parameters for group identity, fixing
* the latent mean and variance at 0 and 1 only in the
* reference group
gsem                                    ///
  (q1_m <- Theta1@a1  _cons@c1) ///
  (q2_m <- Theta1@a2  _cons@c2) ///
  (q3_m <- Theta1@a3  _cons@c3) ///
  (q4_m <- Theta1@a4  _cons@c4) ///
  (q5_m <- Theta1@a5  _cons@c5) ///
  (q6_m <- Theta1@a6  _cons@c6) ///
  (q7_m <- Theta1@a7  _cons@c7) ///
  (q8_m <- Theta1@a8  _cons@c8) ///
  (q9_m <- Theta1@a9  _cons@c9) ///
                                        ///
  (q1_f <- Theta2@a1  _cons@c1) ///
  (q2_f <- Theta2@a2  _cons@c2) ///
  (q3_f <- Theta2@a3  _cons@c3) ///
  (q4_f <- Theta2@a4  _cons@c4) ///
  (q5_f <- Theta2@a5  _cons@c5) ///
  (q6_f <- Theta2@a6  _cons@c6) ///
  (q7_f <- Theta2@a7  _cons@c7) ///
  (q8_f <- Theta2@a8  _cons@c8) ///
  (q9_f <- Theta2@a9  _cons@c9) ///
  ,                                     ///
  logit                                 ///
  variance(Theta1@1 Theta2)             ///
  cov(Theta1*Theta2@0)                  ///
  mean(Theta2)
estimate store m0
```

In this do-file, we make use of the more general Stata model-fitting command for generalized response variables, gsem (see *Stata Structural Equation Modeling Reference Manual* [2017b]). As mentioned earlier in the section, this command is appropriate for the present DIF examination purposes. Further, in the above Stata command file (do-file), we add several annotating comments after an asterisk to enhance comprehensibility. Accordingly, we first assign specific names to the items in either of the groups to subsequently facilitate the constraining of their parameters across groups. Then, for each item in either group, immediately after the gsem command, we state that it evaluates the underlying latent dimension (construct) that is formally denoted Theta1 in the reference group and Theta2 in the focal group. Thereby, the item discrimination index is indicated immediately after that construct name, which is followed by indication of the item difficulty parameter using the reference _cons. [The latter is the c parameter in the intercept-and-slope parameterization in (5.14) in chapter 5 and, along with the item discrimination parameter, completely determines the ICC for any given item, as is readily deduced from that equation. Based on the definition of the c parameter there, the group invariance in it is obviously a necessary and sufficient condition for the group identity of the pertinent item difficulty parameter when the associated item discrimination parameter is group invariant. Thus, for our particular purposes in the rest of this chapter only, we can formally refer to the c parameter for convenience as the difficulty parameter; see next.]

Further, in the above do-file, the sign @ signals constraining across groups. Specifically, the parameter after which it appears is set equal to the parameter with the same label later in the file. In this way, the nine discrimination parameters are constrained for equality across groups, as are the nine difficulty parameters. Toward the end of the do-file, we request that a logistic model (from the class used throughout this book) be fit in both groups by using the `logit` option that is added after the comma. Then, we request with the "variance" line that the trait variance be fixed at 1 in the reference group but not in the focal group. Because the two groups are independent, the formal covariance (parameter) between the trait in the reference group and the trait in the focal group is 0. Thus, we fix that parameter at the latter value in the following "covariance" line. Because the mean of the studied trait is to be also free in the focal group, the last line before `estimate store m0` states only the trait's name in the latter group. The do-file discussed finishes by storing the estimates of this model M0, using the command `estimate store m0`. In in this way, we make its results available for the LRT to be conducted subsequently (see below).

To execute this do-file, all we need to do now is click on the right-most symbol in the toolbar of the Do-file Editor, the line below the one starting with `File`, with the cursor moved over it (that symbol or icon will bring up the text `Execute (do)` attached to the lower end of the cursor). To facilitate carrying out the LRTs described in step 2 above, we need to fit next, one at a time, the $2k = 18$ more general models in this setting. Each of these 18 models has 1 more parameter than M0, namely, the freed discrimination or difficulty parameter. Their do-files are identical to the above after the `gsem` command (and before the `estimate store` command) but with the pertinent parameter not being set equal across the groups. These do-files include as last the command `lrtest m0` (see next for an example).

For instance, to use the LRT for testing whether the discrimination parameter of the first item is the same in both groups, we use the following do-file, which is readily obtained by deleting the symbols @a1 immediately after `Theta1` in the corresponding first line after the `gsem` command in the above discussed do-file (we then highlight only the following part in that original do-file and execute it as described earlier):

```
gsem                                 ///
 (q1_m <- Theta1  _cons@c1)     ///
 (q2_m <- Theta1@a2  _cons@c2)  ///
 (q3_m <- Theta1@a3  _cons@c3)  ///
 (q4_m <- Theta1@a4  _cons@c4)  ///
 (q5_m <- Theta1@a5  _cons@c5)  ///
 (q6_m <- Theta1@a6  _cons@c6)  ///
 (q7_m <- Theta1@a7  _cons@c7)  ///
 (q8_m <- Theta1@a8  _cons@c8)  ///
 (q9_m <- Theta1@a9  _cons@c9)  ///
                                     ///
 (q1_f <- Theta2@a1  _cons@c1)  ///
 (q2_f <- Theta2@a2  _cons@c2)  ///
 (q3_f <- Theta2@a3  _cons@c3)  ///
 (q4_f <- Theta2@a4  _cons@c4)  ///
 (q5_f <- Theta2@a5  _cons@c5)  ///
 (q6_f <- Theta2@a6  _cons@c6)  ///
 (q7_f <- Theta2@a7  _cons@c7)  ///
 (q8_f <- Theta2@a8  _cons@c8)  ///
 (q9_f <- Theta2@a9  _cons@c9)  ///
                                     ///
 ,                                   ///
 logit                               ///
 variance(Theta1@1 Theta2)      ///
 cov(Theta1*Theta2@0)           ///
 mean(Theta2)
lrtest m0
```

The last do-file produces the following output of relevance for our DIF examination purposes here, when testing the group identity in the discrimination parameter of the first item:

```
Likelihood-ratio test                        LR chi2(1)  =      1.24
(Assumption: m0 nested in .)                  Prob > chi2 =    0.2646
```

In this way, after carrying out the remaining 17 tests for individual item discrimination and difficulty parameter identity across groups, we obtain all p-values associated with the tested null hypotheses, which are presented in table 10.1.

Table 10.1. Likelihood-ratio test statistics for the 18 relaxed models, testing discrimination and difficulty parameter identity across groups—with model M0 being nested in each of them—and associated p-values

Parameter	LRT statistic	d.f.	p
a_1	1.24	1	0.265
a_2	1.56	1	0.211
a_3	0.16	1	0.692
a_4	7.77	1	0.005
a_5	7.33	1	0.007
a_6	0.18	1	0.668
a_7	0.05	1	0.826
a_8	5.26	1	0.022
a_9	0.44	1	0.507
b_1	15.56	1	0.000
b_2	1.21	1	0.272
b_3	5.50	1	0.019
b_4	5.45	1	0.029
b_5	5.22	1	0.022
b_6	0.620	1	0.431
b_7	3.67	1	0.055
b_8	3.51	1	0.061
b_9	3.23	1	0.072

Note: Parameter = (single) relaxed parameter across groups, relative to model M0; LRT statistic = LRT statistic value associated with fit relaxed model (with corresponding discrimination or difficulty parameter free across groups), relative to M0; d.f. = degrees of freedom; $p = p$-value associated with the pertinent LRT.

We now wish to find out which, if any, of the 18 null hypotheses of no group differences in an item discrimination or difficulty parameter are to be rejected. As discussed earlier in the chapter, such a hypothesis rejection will indicate DIF with respect to that parameter and item and hence in the overall nine-item test under consideration. To this end, we need to carry out as mentioned multiple testing on the analyzed dataset. To accomplish this, we apply the increasingly popular BH multiple testing procedure (see Benjamini and Hochberg [1995] and Wasserman [2004]; details on the BH procedure, including its rationale, are provided in the Appendix to this chapter).

To commence the BH method application here, we need to evaluate 18 critical values that are defined as follows (see also Appendix for their more general definition in terms of m hypotheses to be tested on a given dataset; $m > 1$):

$$l_j = j\alpha/\{m(1 + 1/2 + 1/3 + \cdots + 1/m)\}$$

$(j = 1, \ldots, 18)$. We obtain these 18 values with Stata using the following short program (type in the command window its first row and press **Enter**, then continue with typing the following row, etc.):

```
. forvalues i = 1/18 {
  2. display `i'*.05/(18*(1+1/2+1/3+1/4+1/5+1/6+1/7+1/8+1/9+1/10+1/11+1/12+1/13+
> 1/14+1/15+1/16+1/17+1/18))
  3. }
```

Stata returns the following l-values in increasing order:

```
.00079476
.00158952
.00238428
.00317905
.00397381
.00476857
.00556333
.00635809
.00715285
.00794762
.00874238
.00953714
.0103319
.01112666
.01192142
.01271619
.01351095
.01430571
```

To use now the BH multiple testing procedure, we can create a Stata data file with the p-values in the last column of table 10.1, sort them in ascending order, and list them for easy comparison correspondingly with the above l-values:

```
. infile p_value using http://www.stata-press.com/data/cirtms/p_values.dat, clear
(18 observations read)

. sort p_value

. list
```

	p_value
1.	0
2.	.005
3.	.007
4.	.019
5.	.022
6.	.022
7.	.029
8.	.055
9.	.061
10.	.072
11.	.211
12.	.265
13.	.272
14.	.431
15.	.507
16.	.668
17.	.692
18.	.826

We next compare each of these 18 p-values, from top to bottom, with the corresponding l-value calculated earlier. That is, we compare the top p-value with the top l-value, the second from top p-value with the second l-value, etc. Stated differently, we compare first the smallest of the 18 p-values in the last presented list, which is 0, with the smallest l-value, that is, 0.001 (rounded off). Then, we compare the second smallest p-value in table 10.1, that is, 0.005, with the second smallest l-value, that is, 0.002. We continue in this way until we find the largest p-value that does not exceed its corresponding l-value (Benjamini and Hochberg 1995). If this happens at the sth step ($1 \leq s \leq 18$), according to the BH procedure, we reject the s null hypotheses associated with the s smallest p-values from the $2k$ tested hypotheses (from the above 18 hypotheses). Alternatively, if all p-values turn out to be larger than their corresponding l-values, we conclude that $s = 0$. In this case, none among the tested null hypotheses should be rejected with this procedure. We stress that in the currently considered DIF examination context, this is the only empirical situation in which we would not conclude that DIF is present for a studied measuring instrument or item set.

In the current empirical example, with the above comparisons of p-values to their corresponding l-values, we easily find out that $s = 1$. This is because each of the rank-ordered p-values in table 10.1 after the smallest one is larger than its corresponding l-value, as worked out above. Therefore, according to the BH procedure, we reject only 1 of the tested 18 null hypotheses. This is the hypothesis associated with a p-value of 0.000 (rounded off), that is, the null hypothesis with the smallest p-value. That is, we reject the hypothesis stating group equality in the difficulty parameter of the first item. We thus identify this item as exhibiting DIF and therefore conclude that the nine-item test under consideration here exhibits DIF.

10.6   Chapter conclusion

In this chapter, we were concerned with an issue in IRT and item response modeling that has received a great deal of attention by methodologists, substantive researchers, and the general public over the past several decades. This is the problem of DIF, which is of special relevance particularly in high-stake testing. We discussed observed variable-based methods for its study, using logistic regression and a related procedure, as well as an IRT-based method (a latent variable-based method) for DIF examination (see also Raykov et al. [Forthcoming] and Osterlind and Everson [2009]). These methods are readily applied with Stata, which offers a flexible approach to the study of DIF and permits the location of items possibly exhibiting DIF in multicomponent measuring instruments or item sets under consideration.

Appendix. The Benjamin–Hochberg multiple testing procedure: A brief introduction

Behavioral, biomedical, educational, clinical, and social scientists are frequently confronted with the necessity to test multiple null hypotheses on a given dataset (for example, Raykov et al. [2017]). A possible approach to accomplishing this simultaneous hypotheses testing activity is to use a conventional multiple testing procedure (MTP) concerned with controlling the familywise error rate (FWER), such as the popular Bonferroni procedure (for example, Agresti and Finlay [2009]). However, the latter is in general conservative, that is, associated with considerable loss in power or diminished sensitivity to incorrect null hypotheses, particularly with a relatively large number of tested hypotheses (for example, Johnson and Wichern [2007]). An alternative approach is based on the false discovery rate (FDR), which is concerned instead with the expected rate of false rejections in a set of tested hypotheses, denoted $H_{0,1}, H_{0,2}, \ldots, H_{0,m}$ ($m > 1$; for example, Benjamini and Hochberg [1995]). Specifically the FDR, designated ϕ, is defined as

$$\phi = \left\{ \begin{array}{l} \mathcal{E}(V/R), \text{ if } R > 0, \text{ and} \\ 0, \text{ if } R = 0 \end{array} \right.$$

where $\mathcal{E}(\cdot)$ symbolizes mean (expectation), V the number of falsely rejected null hypotheses, and R is the overall number of rejected null hypotheses out of the set of k tested. (Obviously, $0 \leq V \leq R \leq m$ holds.)

Benjamini and Hochberg (1995) showed that their FDR-based procedure—referred to as the Benjamin–Hochberg (BH) procedure in the rest of this appendix—not only controls the FWER at the same prespecified significance level α (for example, $\alpha = 0.05$) but also is associated in general with marked gain in power relative to traditional MTPs (compare Raykov, Lichtenberg, and Paulson [2012]). That is, the BH procedure defined in detail below possesses enhanced sensitivity to incorrect null hypotheses compared with conventional MTPs while controlling at the same level the FWER. In addition, the BH procedure addresses an important question that may be considered unattended to by those earlier MTPs. As is well known, the latter control the probability of at least one incorrect rejection from a given set of m assumed correct null hypotheses, that is, the FWER (Agresti and Finlay 2009). However, traditional MTPs do not control (or limit) the number V of such incorrect rejections, whether all of these hypotheses are correct or not, in particular in case at least one null hypothesis is incorrect to begin with. In contrast to those conventional methods, the BH procedure limits—that is, keeps small, or controls—the expected rate of incorrect rejections. This is achieved by effectively precluding the number V of incorrectly rejected null hypotheses from becoming (on average) large relative to the overall number R of rejected hypotheses out of the m tested (see below).

The BH procedure can be applied when there are $m > 1$ hypotheses to be tested for a given dataset (with regard to one or more types of parameters). The procedure consists of the following steps (compare Wasserman [2004]). Denote first by $p_1, p_2, \ldots, p_m$ the p-values of the m hypotheses, and let $p_{(1)} \leq p_{(2)} \leq \ldots \leq p_{(m)}$ be their sequence in ascending order. For a prespecified significance level α (for example, $\alpha = 0.05$), define next the following series of m ratios (referred to as l-values):

$$l_j = j\alpha / \{m(1 + 1/2 + 1/3 + \cdots + 1/m)\} \quad (j = 1, \ldots, m)$$

Next, let r be the largest number j for which the inequality

$$p_{(j)} \leq l_j \tag{A.1}$$

holds; alternatively, if (A.1) does not hold for any j, define $r = 0$ and $p_{(0)} = 0$. This p-value $p_{(r)}$ is often called the BH rejection threshold and denoted T here (that is, $T = p_{(r)}$). Then, at the decision stage, the BH procedure declares as rejected all null hypotheses with p-values that do not exceed $p_{(r)}$, that is, all null hypotheses with p-values smaller or equal to the threshold T; thereby, if $r = 0$, none of the tested p null hypotheses are to be rejected.

Benjamini and Hochberg (1995) and Benjamini and Yekutieli (2001) demonstrated that regardless of i) how many of the m null hypotheses are actually true and ii) how their associated p-values are distributed,

$$\phi \leq \alpha$$

always holds. That is, the BH procedure controls the number of incorrect "discoveries" through limiting its expected rate by a fairly small number (for example, 0.05). Those authors also demonstrated that the BH method controls the FWER by the same α while possessing in general higher power than an MTP controlling the FWER at that α level. This advantage of the BH procedure over traditional MTPs increases with the number of incorrect null hypotheses among the set of m hypotheses tested.

11 Polytomous item response models and hybrid models

In the preceding chapters of the book, we have dealt with binary response items or binary scored items. A main reason for our focus on them was the realization that this type of item response or scoring covers a great deal of empirical applications of item response theory (IRT) in the behavioral, educational, and social sciences. At the same time, however, an increasingly larger field of IRT utilization in these and cognate disciplines, including the clinical, social, organizational and communication sciences, as well as business and marketing, frequently uses items that have more than two answer options. Such items are typically referred to as polytomous items. IRT models for polytomous items, fittingly called polytomous IRT models, are therefore the main concern of the present chapter. In it, we still remain within the framework of unidimensional IRT as indicated earlier. (We discuss multidimensional IRT in the following chapter.)

11.1 Why do we need polytomous items?

There are several important reasons why polytomous IRT models are enjoying growing popularity in the social, clinical, and business sciences (for example, de Ayala [2009]). One of the most important reasons may well be that in these and cognate disciplines, it is increasingly more often of interest to evaluate a respondent's typical performance rather than his or her maximal performance. It is the latter type of performance, however, for which binary or binary scored items (dichotomous items) tend to be more appropriate and perhaps best used, as indicated earlier in the book. Yet when studying typical performance, references to "correct" (true, right) response tend to lose their common meaning. In fact, there is frequently no such response to the majority of items, if not all items, in an overall measure, item set, or multi-item instrument for evaluating this type of subject performance.

A related reason is that dichotomous distinctions, as in binary or so scored items, are less clear in certain areas of measurement—like those indicated above—than are polytomous distinctions for items with more than two possible answers (for example, Nering and Ostini [2010]). Indeed, dichotomous items may be rather limited in terms of the extent of information they contain or provide in the end about underlying latent dimensions being evaluated, such as certain traits, opinions, proficiency, mastery, aptitudes, or attitudes. Similarly, in many areas in the social sciences, more subtle nuances of agreement or disagreement are needed for measurement of these kinds of latent dimensions than could be achieved by using only (or even mostly) binary items.

Because of these and related reasons, measurement items with several possible response options are increasingly more widely used in the behavioral, educational, and social sciences and well beyond them. This is a trend that has been notable across these and cognate disciplines starting perhaps as early as the 1960s. For these reasons, the following have become highly popular types of polytomous items in empirical studies in these and cognate disciplines (for example, de Ayala [2009]):

i) rating scale items, such as the widely used Likert-type items, especially in the behavioral, clinical, and social disciplines;

ii) ability test items that provide partial credit for partially correct answers (for example, items used in educational assessments, for instance, portfolio assessment test formats); and

iii) multiple-choice items where each response option is scored separately.

The characteristic feature of polytomous items is that they can provide more information about differences among studied individuals than dichotomous measures (for example, Nering and Ostini [2010]). In addition, these differences are captured across a wider range of the trait, construct, or ability continuum of interest than is the case with dichotomous items. Thus, polytomous items provide in general more statistical (statistically relevant) information. This leads eventually to more precise individual trait, construct, or ability level estimates than in the case of binary or binary scored items. Therefore, the present chapter is concerned with IRT models that accommodate polytomous items.[1]

11.2   A key distinction between item response theory models with polytomous and dichotomous items

A major difference between polytomous IRT models and those for binary (scored) items lies in the following fact (for example, de Ayala [2009]). For dichotomous items, as discussed in chapter 2, knowledge of the item characteristic curve (ICC) for the "correct" response (that is, with score "1") completely determines the ICC for "incorrect" response (with score "0"). Indeed, as indicated earlier, this is achieved through simple subtraction from 1: the ICC for the response "0" is obtained as the complement to 1 of the ICC for the response "1" (see chapter 2). However, this exact relationship is no longer true

---

1. While a binary or binary scored (dichotomous) item could in principle be considered a special case of a polytomous item with r response options ($r \geq 2$), namely, when $r = 2$, we will reserve the term "polytomous item" for the case $r > 2$ throughout this chapter. Correspondingly, the term polytomous IRT model will be reserved for models with such items. Alternatively, for items with two response options, we will use the term "binary/binary scored item", or "dichotomous item", as in previous chapters of the book and chapter 12. Relatedly, we emphasize that in this chapter, we consider the response options to be denoted "1", "2", ..., "r" for an item under consideration, unless specifically indicated otherwise (namely, as "0", "1", ..., "r" in some other polytomous item and model treatments in the IRT literature).

for polytomous items. In fact, each category response function (CRF) for a polytomous item needs to be structured or modeled separately (see below; a CRF is at times also called an "item trace line" or a "category characteristic curve", CCC.)

This observation leads us to the following important definition of a counterpart concept here to that of ICC for dichotomous items. A CRF for a given response option (category, or answer choice) of a polytomous item is defined as the probability for response in that category as a function of the underlying trait, construct, or ability, θ. (As mentioned above, θ is assumed to be unidimensional in this chapter.) We note that the ICC for a binary (scored) item is a special case of CRF, namely, when only two response options are present as in a dichotomous item. Conversely, the CRF is a generalization of the ICC to the case with more than two possible response options. Furthermore, an implied distinction between the ICC and CRF is that the CRFs for polytomous items are not exclusively monotonic as in the binary case (see chapter 2). In particular, for items with ordered categories, only the functions for the extreme negative (lowest) and extreme positive (highest) categories are monotonically decreasing and increasing, respectively. Such nonmonotonic functions, as some CRFs are for a polytomous item, present special problems to handle within a model for these types of items. In fact, these functions can no longer be described in relatively simple terms and parameters using, say, a single scale (slope) and location (intercept) parameter. However, the latter simplicity is a characteristic feature, for instance, of the two-parameter logistic (2PL) model with dichotomous items. Indeed, as elaborated in chapters 4 and 5, the ICCs in the latter model are readily representable in terms of their corresponding a and b parameters (item discrimination and difficulty parameters, as scale and location parameters, respectively).

With the preceding discussion in mind, in the rest of this chapter, we will be concerned with several widely used polytomous IRT models and will illustrate their empirical application using Stata.

11.3   The nominal response model

One of the first models proposed for items with more than two response options is the nominal response model (NRM). This model goes back to an influential paper by Bock (1972). The NRM does not assume any ordering of the responses on each of a given set of items, with these responses being at least two on any item. The model may be seen as a generalization of the 2PL model for the case with more than two possible answers, as in a polytomous (unordered category) test item. Hence, the NRM can also be viewed as a generalization of the Rasch model, because the latter is a special case of the 2PL model (see chapter 6).

To define the NRM, suppose that for the jth item from a unidimensional set with k items of interest (for example, a multicomponent measuring instrument) there are r unordered response options ($1 \le j \le k; k, r > 1$). That is, the response options on this item represent a set of r mutually exclusive, exhaustive, and unordered (nominal) categories. In such items, the score formally assigned to each category—such as, say, $1, 2, \ldots$, or r—has no meaning per se other than to designate a corresponding response option. (Thus, we will use in the remainder of this chapter the term "numeral" to refer to a score used to designate a response on a polytomous item; see, for example, Michell [2005].) The NRM assumes then that the probability for an examined person to select option q, denoted $P_{jq}(\theta)$, is as follows (θ denotes as usual his or her level of the trait, construct, or ability of interest to evaluate),

$$P_{jq}(\theta) = \exp\{a_{jq}(\theta - b_{jq})\} / \sum_{h=1}^{r} \exp\{a_{jh}(\theta - b_{jh})\} \tag{11.1}$$

where for each θ

$$P_{j1}(\theta) + \cdots + P_{jr}(\theta) = 1 \tag{11.2}$$

holds ($j = 1, \ldots, k, q = 1, \ldots, r$). We note that (11.2) is valid not only for the NRM discussed here but also for each model considered in the present chapter. This is because (11.2) represents the unit sum of the probabilities of all possible response options for the jth item—a relationship that should obviously hold at any level of the underlying ability or trait $\theta(1 \le j \le k)$. (In the rest of this chapter, unless otherwise stated, we assume as indicated above that the r response options for a given item are denoted $1, 2, \ldots, r$.)

We similarly notice from (11.1) and (11.2) that it is the pair of parameters (a_{jq}, b_{jq}), rather than a single parameter, that characterizes the qth response option of the item in question ($q = 1, \ldots, r; j = 1, \ldots, k$). Analogous to the 2PL model, a_{jq} is called a discrimination parameter and b_{jq} difficulty parameter for the jth item (see next). However, we note that unlike the case with that earlier discussed model, there is a pair of subindexes here that are of relevance and attached to either of these two parameters for any item following the NRM. In particular, this pair includes a subindex indicating the category to which the associated parameter belongs, in addition to that subindex for item. Thus, a_{jq} is more precisely referred to as the discrimination parameter of category q, while b_{jq} is the difficulty parameter of that category of the jth item ($q = 1, \ldots, r; j = 1, \ldots, k$).

As could be expected based on our discussion in chapters 4 and 5 on the relationship between IRT and logistic regression, when we interpret the results obtained with the NRM, it is useful to recall the way one deals with and interprets the multinomial regression model and associated results (Agresti 2013). The reason is that the latter model can be seen as the analogue to the NRM in the observed predictor case (see also chapter 4). In particular, it is important in empirical applications of the NRM to be aware of which category is to be treated as a "reference" (baseline) category. In this connection, we note that Stata assigns by default the lowest numbered response as a "base" or "reference" category, for example, the one designated by the numeral "1" (when smallest among all other numerals used to denote response categories). This implies also that by definition, $a_{j1} = 0$ and $b_{j1} = 0$ for model identification ($j = 1, \ldots, k$). Therefore, the discrimination and difficulty parameters of the other categories, namely, a_{jq} and b_{jq}, can be interpreted, respectively, as discrimination and difficulty parameters associated with the choice or preference of the qth category relative to (versus) that base or reference category for the jth item ($q = 2, \ldots, r; j = 1, \ldots, k$). Furthermore, revisiting footnote 1 to chapter 5, we see this item and category discrimination parameter, b_{jq}, also as reflecting the propensity to choose the qth response category instead of the baseline category on the jth item ($q = 2, \ldots, r; j = 1, \ldots, k$).

With the above setting of vanishing discrimination and difficulty parameters for the reference category, from (11.1) and (11.2) when $r = 2$, one readily obtains the ICCs of the 2PL model for the jth item:

$$P_{j1}(\theta) = 1/\left[1 + \exp\left\{a_{j2}(\theta - b_{j2})\right\}\right], \text{and}$$
$$P_{j2}(\theta) = \exp\{a_{j2}(\theta - b_{j2})\}/[1 + \exp\{a_{j2}(\theta - b_{j2})\}]$$

The last two are in fact the defining equations of the 2PL model for a dichotomous item, if treating the reference category as being assigned the "score" 0 and the second (other) category as being assigned the "score" 1, $j = 1, \ldots, k$ (see chapter 6). (We note that formally in the last two equations, we do not need the second subindex to a and b for a given item, because there is only one nonredundant ICC then, say, for the category "1": the ICC for the other category is the complement to 1 of the former ICC; see also chapter 2.) This discussion shows formally that the NRM is a generalization of the 2PL model, and thus of the Rasch model as well, to the polytomous unordered item response case. Conversely, the 2PL model is a special case of the NRM when $r = 2$, and the Rasch model yet a further special case of the latter, with item-invariant discrimination parameters.

11.3.1   An empirical illustration of the nominal response model

For our particular aims here, we will use data obtained from $n = 514$ patients on $k = 5$ items from a scale measuring presumably anxiety, with the items presenting statements about feelings with respect to certain events in one's life. (The original data can be downloaded directly from http://www.ssicentral.com, with the items in it asking specifically about feeling calm, tense, regretful, at ease, and nervous. For the illustration purposes in the rest of this chapter, we refer to the construct measured by these five items as anxiety.) The data on these polytomous items are found in `anxiety5items.dta`. The following answer options were available to the studied persons responding to their feelings with respect to life events: "not at all applicable to me", "rarely applicable to me", "sometimes applicable to me", "often applicable to me", and "always applicable to me". These responses are denoted or coded in the data file by using correspondingly the numerals 1 through 5.

To demonstrate only an application of the NRM, for the sake of the present example, let us consider these answer options treated as unordered (see later in this chapter for extensions). We thus read in the data file first and as usual take a look at the variables in it:

```
. use http://www.stata-press.com/data/cirtms/anxiety5items.dta
```

```
. describe

Contains data from http://www.stata-press.com/data/cirtms/anxiety5items.dta
  obs:           514
  vars:            5                              12 Oct 2016 14:42
  size:       10,280
```

variable name	storage type	display format	value label	variable label
calm	float	%9.0g		
tense	float	%9.0g		
regretful	float	%9.0g		
atease	float	%9.0g		
nervous	float	%9.0g		

```
Sorted by:
```

We are now ready to fit the NRM model, which we achieve with the following Stata command:

```
. irt nrm calm-nervous

Fitting fixed-effects model:

Iteration 0:   log likelihood = -3599.4342
Iteration 1:   log likelihood = -3599.4342

Fitting full model:

Iteration 0:   log likelihood = -3463.3374  (not concave)
Iteration 1:   log likelihood = -3327.5507
Iteration 2:   log likelihood = -3198.1987
Iteration 3:   log likelihood = -3180.7607
Iteration 4:   log likelihood = -3179.5186
Iteration 5:   log likelihood =  -3179.511
Iteration 6:   log likelihood =  -3179.511

Nominal response model                      Number of obs     =      514
Log likelihood = -3179.511
```

	Coef.	Std. Err.	z	P>|z|	[95% Conf. Interval]	
calm						
Discrim						
2 vs 1	2.058268	.3737465	5.51	0.000	1.325739	2.790798
3 vs 1	3.836122	.5204513	7.37	0.000	2.816057	4.856188
4 vs 1	6.537603	.8317523	7.86	0.000	4.907399	8.167808
5 vs 1	6.854232	1.036193	6.61	0.000	4.82333	8.885133
Diff						
2 vs 1	-.884001	.0960933	-9.20	0.000	-1.07234	-.6956616
3 vs 1	-.3058478	.0797372	-3.84	0.000	-.4621298	-.1495659
4 vs 1	.4052133	.1073355	3.78	0.000	.1948397	.6155869
5 vs 1	.6975946	.1398809	4.99	0.000	.4234332	.9717561

tense						
Discrim						
2 vs 1	1.741124	.3219806	5.41	0.000	1.110053	2.372194
3 vs 1	3.051999	.4373235	6.98	0.000	2.194861	3.909137
4 vs 1	4.729577	.5672034	8.34	0.000	3.617879	5.841276
5 vs 1	6.005388	.7925817	7.58	0.000	4.451956	7.558819
Diff						
2 vs 1	-.9630793	.1099569	-8.76	0.000	-1.178591	-.7475677
3 vs 1	-.4077822	.0836613	-4.87	0.000	-.5717554	-.243809
4 vs 1	.0930578	.0919089	1.01	0.311	-.0870803	.2731958
5 vs 1	.6292221	.1287067	4.89	0.000	.3769615	.8814826
regretful						
Discrim						
2 vs 1	.9108159	.175889	5.18	0.000	.5660799	1.255552
3 vs 1	1.377087	.2223204	6.19	0.000	.9413466	1.812827
4 vs 1	2.167399	.2824531	7.67	0.000	1.613801	2.720997
5 vs 1	2.940892	.4387327	6.70	0.000	2.080992	3.800792
Diff						
2 vs 1	-.4240454	.1372531	-3.09	0.002	-.6930565	-.1550344
3 vs 1	.2181196	.127561	1.71	0.087	-.0318953	.4681346
4 vs 1	.5647008	.1120826	5.04	0.000	.3450229	.7843786
5 vs 1	1.122213	.1362012	8.24	0.000	.8552633	1.389162
atease						
Discrim						
2 vs 1	2.889374	.6571248	4.40	0.000	1.601433	4.177315
3 vs 1	4.978993	.8102032	6.15	0.000	3.391024	6.566962
4 vs 1	6.940502	.9260655	7.49	0.000	5.125447	8.755557
5 vs 1	8.240645	1.167154	7.06	0.000	5.953066	10.52822
Diff						
2 vs 1	-1.078142	.0961554	-11.21	0.000	-1.266603	-.8896811
3 vs 1	-.5637687	.083228	-6.77	0.000	-.7268927	-.4006448
4 vs 1	-.0397348	.1102134	-0.36	0.718	-.2557491	.1762794
5 vs 1	.4672956	.1707959	2.74	0.006	.1325417	.8020494
nervous						
Discrim						
2 vs 1	1.093094	.2157188	5.07	0.000	.6702933	1.515896
3 vs 1	1.930749	.2737273	7.05	0.000	1.394254	2.467245
4 vs 1	2.593354	.3224593	8.04	0.000	1.961345	3.225363
5 vs 1	3.362253	.4445859	7.56	0.000	2.490881	4.233625
Diff						
2 vs 1	-.6077147	.1272378	-4.78	0.000	-.8570963	-.3583331
3 vs 1	-.080439	.0977557	-0.82	0.411	-.2720367	.1111587
4 vs 1	.2548234	.0988217	2.58	0.010	.0611364	.4485103
5 vs 1	.7955258	.1246966	6.38	0.000	.5511249	1.039927

Before we discuss these results, let us also obtain the CRFs (or CCCs) for each of the five items analyzed. To this end, we use sequentially the next command with each of them in turn, first for the initial two items and further below for the remaining three items (note the use of the keyword `icc` while requesting the CCCs):

```
. irtgraph icc calm
```

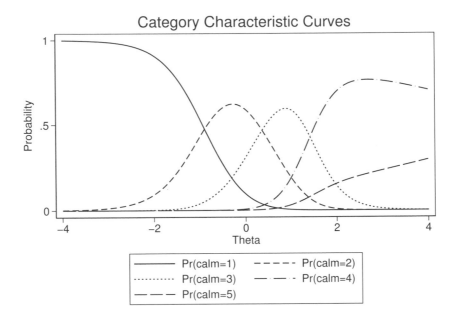

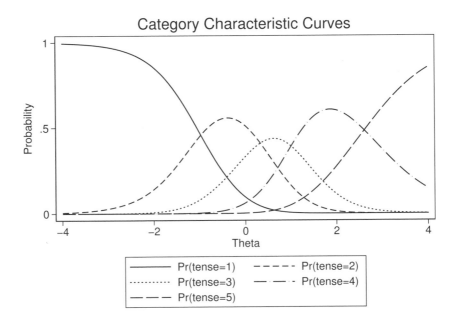

We readily observe from these two plots the earlier mentioned characteristic feature of polytomous IRT models, including the NRM (see section 11.2). Specifically, apart from the categories denoted with the lowest and highest numerals (here 1 and 5), all other CCCs are not monotonic functions of the underlying latent trait, θ, evaluated by this scale. Thereby, as can be expected, the CRF of the lowest category ("not at all applicable to me") is monotonically decreasing and "shifted" markedly to the left, that is, toward the lower end of the θ-scale. At the same time, the CCC for the highest denoted response option ("always applicable to me") is monotonically increasing and "shifted" to the right of the presented range on that scale. Further, as follows from (11.2) (and indicated earlier in the chapter), at any point on the underlying latent dimension, the sum of the probabilities across all answer options is 1. That is, erecting at any point on the horizontal axis a vertical line in either of the above 2 or of the following 3 graphs—as well as those associated with all polytomous models considered in this chapter—and then adding the heights at which it intersects the CRFs of the pertinent item should yield 1.

Moreover, for either of these two items ("feeling calm" and "feeling tense", as well as for the remaining three items), we notice an important pattern. Specifically, their CRFs are located further to the right along the latent dimension as the frequency increases of the occurrence of the feeling asked about in the item (that is, as moving from the "not applicable to me" through the "always applicable to me" response category). This suggests that each of these two items may actually be functioning as an ordinal measure. We will further explore this observation later in the chapter when using ordinal polytomous models on this dataset as well (see also section 11.7 below for additional discussion).

The CRFs for the other three items follow next, which are obtained with Stata as illustrated above (only changing the name of the item at the end of the last used Stata command); they effectively corroborate the last observation (see preceding paragraph, and note the legend at the bottom of the graph below that contains the item name).

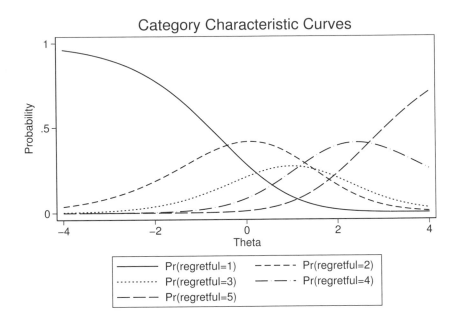

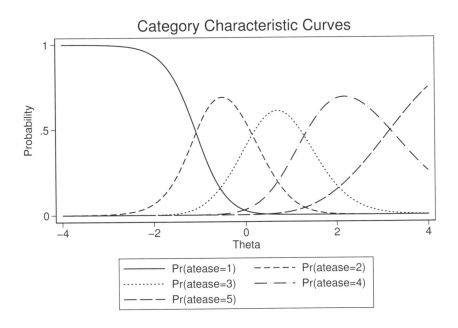

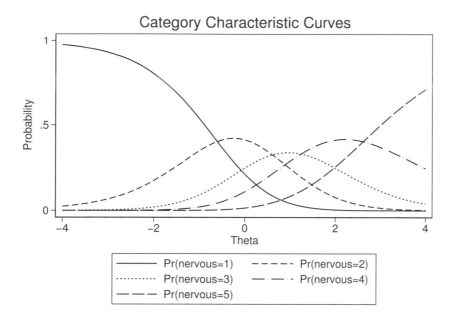

As implied from the earlier NRM discussion, for each of these five items, the crossing point of the CRF of any category after the first, with the CRF of the base response option, reflects the difficulty parameter associated with that category. This is the point on the horizontal axis (the θ-scale), above which the probability of choosing this category becomes larger than that for the reference (base) response. For instance, looking at the output panel pertaining to the item "feeling calm" (the first item listed), we find that for $\theta > -0.88$ (rounded off) the probability of choosing the response option 2 becomes for the first time higher than choosing the reference category (response option 1) as one moves from left to right on the horizontal axis, that is, the θ-scale. Conversely, the probability for choosing response 2 is lower before that point on the trait dimension, that is, for persons with $\theta < -0.88$. Similarly, someone with $\theta = -0.88$ (rounded off) has the same probability of choosing category 1 as he or she does for selecting category 2 (see also below). Looking next at the graph with CCCs for the item "feeling tense" and inspecting the intersection points of its associated CCC, we note that persons with $\theta > 2.6$ (approximately) have the highest probability of choosing category 5 relative to any other category. (This is the position on the θ-scale where the CRFs of the highest two response categories for this item intersect.) The remaining items' category difficulty parameters and CCC intersection points are interpreted along the same lines.

Further, the item discrimination parameters are interpretable as reflecting the gradient (steepness in the "central part") of the ICC when juxtaposing the first with each subsequent category in turn for a given item. For instance, for the "nervous" item, the discrimination parameter for the last category is highest (namely, estimated at 3.36), as could be expected because of the implied largest discrepancy between it and the first or lowest category on this item. More generally, looking at (11.1), which is an increasing function in the a parameter for a given item and option, and recalling that we assumed positive discrimination parameters throughout (see chapter 5), we observe the following relationship. For a given increment in θ, a higher increase in the associated probability of choosing a particular item category versus the baseline (reference or base) category is ensured for the item and response option associated with a higher discrimination parameter, all else being the same.

Returning for a moment to the above interpretation of the item category difficulty parameters, we can further illustrate that discussion by drawing the CCCs for, say, the `calm` item while highlighting its estimated category difficulty parameters. We achieve this with the following Stata command, with the plot presented after it (notice the use in the subcommand of the item category difficulty parameter estimates, rounded off, from the last presented Stata output):

```
. irtgraph icc calm, xlabel(-4 -.88 -.3 .4 .7 4, grid alt)
```

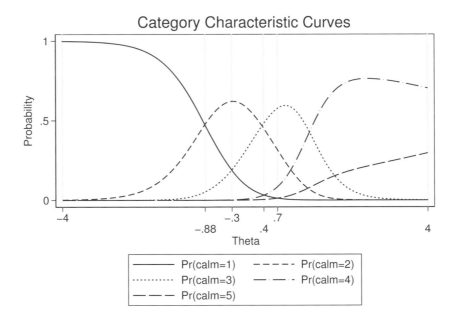

In this graph, we use as boundaries on the horizontal axis the numbers -4 and 4 as before (see chapter 2). The reason is that under normality, a z score (which a θ-value actually is, given the standardization used for model identification) is highly unlikely to be larger than 4 in absolute value (for example, Raykov and Marcoulides [2013]). We also notice from this graph that for trait levels below (-0.88), the base category (with response denoted by the numeral "1" in the data, or first category) is associated with the highest probability for being chosen, and for most of this range on the θ-scale, that probability is higher than the sum of those for choosing the remaining categories. In other words, the probability of choosing the first category is higher there than that for choosing all other categories combined. Further, from that point on the θ-scale up until 0.4 (approximately), the most probable category is that associated with the numeral "2" (second category). From the last mentioned point on this scale up until approximately 1.5 (that is, 1.5 standard deviations above the mean of θ), the most probable category is the third, and so on. The corresponding quantities for the other four items are interpreted in the same way.

To furnish the item information functions (IIFs) of the five items, we use the following Stata command (see below for their resulting joint plot):

```
. irtgraph iif
```

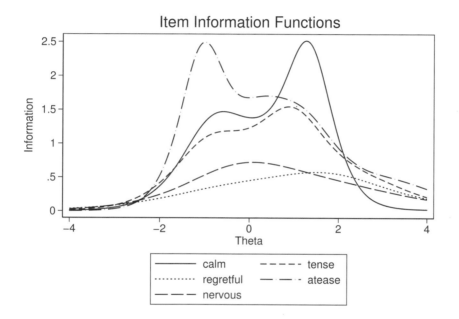

We readily see from this graph that the `calm` item provides the highest information about a location on the underlying anxiety continuum (dimension) between approximately 1 and 2 standard deviations above its mean. This information is markedly above that in the "at ease" item with respect to the same latent continuum range and is also notably higher there than the information associated with any of the other three items. We can interpret along similar lines the remaining item differences with respect to information provided about the studied latent dimension (in corresponding regions on it).

We also easily notice from the last plot that the IIFs are not unimodal or symmetric. In addition, the IIFs achieve their maxima for values of θ that are located at different positions for different items. The reason is that when we consider polytomous items, as in this example, for each of them, the item response categories may contribute different amounts of information about positions on the underlying latent continuum (Samejima 2016). These amounts of information also differ by item, rendering in the end item-specific shapes of the IIFs, as observed from the last presented plot.

To obtain the overall test information function, that is, for the whole scale consisting of these five anxiety items, we use this command (resulting plot follows it):

```
. irtgraph tif
```

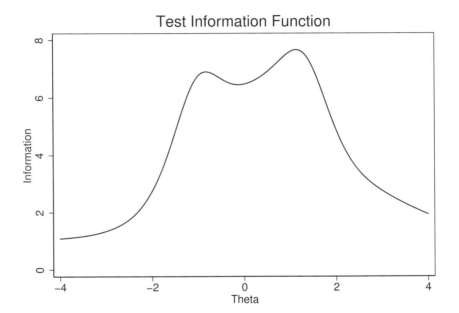

We observe from the test information function that this anxiety scale furnishes the highest information (precision of estimation) about a location around 1.5 standard deviations above the average level of anxiety. In fact, this scale functions about equally well in this respect across the range from approximately 1 standard deviation below the mean on anxiety up to (about) that location with maximal information function value. Moreover, this overall instrument affords substantially less information (precision of estimation) about positions further away from the mean in either direction. In particular, the information this 5-item instrument provides is substantially lower for locations at least 2 standard deviations (approximately) away from the mean on anxiety.

As indicated earlier, during the result interpretations, we have noticed some potential evidence in favor of a hypothesis that the items in this scale function possibly as ordinal measures of the presumed construct of anxiety. We further explore this possibility with the following popular polytomous models that are discussed in the next three sections.

11.4   The partial credit and the rating scale models

Response formats that contain categories meant to be ordered are quite common in the behavioral, educational, and social sciences. When we consider items with answer categories that could be viewed or hypothesized as ordinal, a number of possible polytomous IRT models can be used. Two noteworthy models are the partial credit and the rating scale models (RSMs). Both of them were originally developed nearly four decades ago as extensions of the Rasch model to the ordinal item response category case and are discussed in turn in this section.

11.4.1   Partial credit model

The partial credit model (PCM), often referred to in the literature by its abbreviation, was proposed by Masters (1982). The idea underlying the PCM consists of i) successive dichotomization of adjacent categories for an ordinal item with more than two possible response options and ii) applying the Rasch model for each such dichotomization, allowing only a separate location parameter. To formally define the PCM, we assume that the probability for selecting the qth response option of the jth item in a given set is as follows [for simplicity and to emphasize the relation to the Rasch model for dichotomous items, the discrimination parameter is absorbed next into the individual trait level and difficulty parameter or, alternatively, can be assumed to be 1, as mentioned earlier in the book for the one-parameter logistic (1PL) model],

$$P_{jq}(\theta) = \exp\left\{\sum_{s=0}^{q}(\theta - b_{js})\right\} \Big/ \left[1 + \sum_{h=1}^{r}\exp\left\{\sum_{s=1}^{h}(\theta - b_{js})\right\}\right] \qquad (11.3)$$

$[j = 1, \ldots, k;\ q = 0, 1, \ldots, r;\ r \geq 1;$ for $q = 0$, the numerator in the right-hand side of (11.3) is defined as 1]. In (11.3), b_{jq} can be thought of as a difficulty parameter when distinguishing the sth from the remaining response options (see below for more detail and an example). From (11.3), it can be found that the PCM is a generalization of the Rasch (1PL) model to the case when each item has more than two response options; conversely, the Rasch model is a special case of the PCM with two such options. (Note that, as in the preceding section, we need in the latter case only the form of the ICC for one response option, say, that denoted "1", because the ICC for the other is its complement to 1.)

When is the PCM a viable option for a researcher to pursue? This model is considered appropriate for settings where each of a set of items under consideration requires successive completion of a number of "elementary" tasks, whose correct solution is needed to respond to the item in the right way (for example, de Ayala [2009]). We stress, however, that by its definition, the PCM is not restricted to such "component analytic items". In fact, it can be applied with any item format possessing a finite set of response options, if one can reasonably conceptualize its category boundaries as successive steps that must be "cleared" (completed or accomplished) in turn for the respondents to "locate" themselves in or choose a particular response category.

To fit the PCM with Stata to the anxiety dataset used in the preceding section, we use this command (with results presented following it):

```
. irt pcm calm-nervous

Fitting fixed-effects model:

Iteration 0:   log likelihood = -3599.4342
Iteration 1:   log likelihood = -3599.4342

Fitting full model:

Iteration 0:   log likelihood = -3378.8602
Iteration 1:   log likelihood = -3229.3073
Iteration 2:   log likelihood = -3217.4216
Iteration 3:   log likelihood = -3217.3189
Iteration 4:   log likelihood = -3217.3188

Partial credit model                          Number of obs    =     514
Log likelihood = -3217.3188
```

	Coef.	Std. Err.	z	P>|z|	[95% Conf. Interval]	
Discrim	1.274144	.0677965	18.79	0.000	1.141265	1.407023
calm						
Diff						
2 vs 1	-.9941964	.1160514	-8.57	0.000	-1.221653	-.7667399
3 vs 2	.4435377	.1043609	4.25	0.000	.238994	.6480814
4 vs 3	1.699897	.1602341	10.61	0.000	1.385844	2.013951
5 vs 4	2.750533	.3218528	8.55	0.000	2.119713	3.381353
tense						
Diff						
2 vs 1	-1.059846	.1199974	-8.83	0.000	-1.295037	-.8246554
3 vs 2	.352751	.1072738	3.29	0.001	.1424982	.5630039
4 vs 3	1.075543	.1332424	8.07	0.000	.8143931	1.336694
5 vs 4	2.517478	.2501667	10.06	0.000	2.027161	3.007796
regretful						
Diff						
2 vs 1	-.4894244	.1051203	-4.66	0.000	-.6954564	-.2833923
3 vs 2	.7068919	.1179094	6.00	0.000	.4757938	.9379901
4 vs 3	1.147041	.148435	7.73	0.000	.8561139	1.437968
5 vs 4	2.290527	.2425403	9.44	0.000	1.815157	2.765897
atease						
Diff						
2 vs 1	-1.330473	.1297694	-10.25	0.000	-1.584816	-1.07613
3 vs 2	.1926188	.1014066	1.90	0.058	-.0061345	.3913721
4 vs 3	1.428635	.1405203	10.17	0.000	1.15322	1.70405
5 vs 4	3.039799	.3321829	9.15	0.000	2.388733	3.690866
nervous						
Diff						
2 vs 1	-.6415196	.1102714	-5.82	0.000	-.8576477	-.4253916
3 vs 2	.4391702	.1128415	3.89	0.000	.218005	.6603354
4 vs 3	1.016633	.1369625	7.42	0.000	.7481914	1.285074
5 vs 4	2.196408	.2219737	9.89	0.000	1.761347	2.631468

We note from the output that, as can also be seen from the PCM's defining (11.3), the items share a common discrimination parameter. This constant discrimination parameter assumption will be relaxed in the model considered in the next section, the generalized partial credit model (GPCM). Further, when comparing higher categories with those immediately preceding them (in terms of the used numerals to denote these responses), we see that the difficulty parameters increase for all items and that their values show some similarity across items—when taking into account their standard errors as well. These parameters can be interpreted in the PCM as "thresholds" on the underlying latent dimension being evaluated by the items, such that for a trait level above them, a choice for the higher category is likelier to be made on the pertinent item (compare Masters [2016]). For instance, consistent with this interpretation, looking at the "nervous" item, we see that the confidence intervals of its difficulty parameters for higher categories are entirely above those for the lower categories. This finding suggests possible upward progression of these parameters in the population. Moreover, looking at the first difficulty parameter for the same item, a person with $\theta = -0.64$ (rounded off) has a trait level at which a choice for category "1" (in this example, "not at all applicable to me") is equally probable as that for category "2" ("rarely applicable to me"). At the same time, a person with $\theta = 0.44$ (rounded off) is equally probable to choose either the latter category or category "3" ("sometimes applicable to me"), etc. We observe that in contrast to the NRM, within an item, each category after the first is compared with the one immediately below it. This fact is consistent with the presumably ordinal nature of the possible responses on the item. (See below for further discussion on this matter.)

We can readily graph as follows the CRFs for the "nervous" item and also request vertical lines erected at its above estimated difficulty parameters (in terms of syntax, note that we are still asking for `icc`, as in the dichotomous item case, even though the requested plots are actually of CRFs or CCCs):

```
. irtgraph icc nervous, xlabel(-4 -.64 .44 1.02 2.2 4, grid)
```

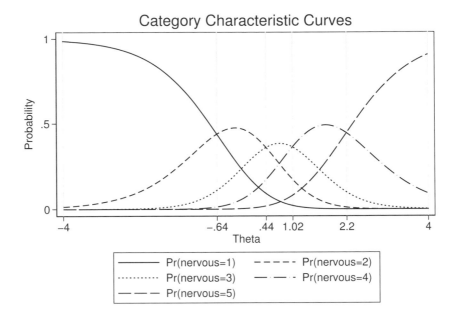

This command produces the plot of the CCCs for the item in question, which is presented above (see preceding discussion for interpretation of the highlighted CCC crossing points by its grid). We notice from this graph, upon examining also the last presented output for the fit PCM, that the estimated difficulty parameters (rounded off) associated with this item represent the crossing points of the CRF of each response category—starting with the first—with the CRF of the one immediately above it. Even though there are more crossing points of CRFs on the last plot, the difficulty parameters are reflected only in those for the pairs of CCCs pertaining to successive response options, as parameterized by the PCM and discussed above in this subsection [see (11.3)].

11.4.2   Rating scale model

The RSM was proposed by Andrich (1978) and, while closely related, represents a more parsimonious version of the PCM. A particular requirement for applicability of the RSM is that all items in a given set or measuring instrument under consideration have the same number of categories. In that case, the RSM can be viewed as a constrained PCM. Specifically, the RSM is a PCM where in addition to the above probability for category choice in (11.3), the constraint of identical discrepancy in any pair of successive category difficulty parameters is imposed across all items. That is, in the RSM,

$$b_{js} - b_{j,s-1} = b_{l,s} - b_{l,s-1} \qquad (11.4)$$

holds for all item pairs and categories $(j, l = 1, \ldots, k; s = 0, 1, \ldots, r; j \neq l)$.

Because of the parameter constraint in (11.4), the RSM is appropriate to consider in empirical settings with ordinal items possessing the property that the same (denoted or coded) responses have the same meaning across items. As implied from this discussion, the defining equations of the RSM in terms of probability of a particular response option, as a function of θ, consist of (11.3) that are augmented with (11.4).

To fit the RSM with Stata to the anxiety dataset used earlier in the chapter, we use this command (with results presented following it):

```
. irt rsm calm-nervous

Fitting fixed-effects model:

Iteration 0:   log likelihood = -3854.2555
Iteration 1:   log likelihood = -3635.9861
Iteration 2:   log likelihood = -3628.1403
Iteration 3:   log likelihood = -3628.0684
Iteration 4:   log likelihood = -3628.0683

Fitting full model:

Iteration 0:   log likelihood = -3411.1862
Iteration 1:   log likelihood = -3264.2465
Iteration 2:   log likelihood = -3251.4182
Iteration 3:   log likelihood = -3251.2556
Iteration 4:   log likelihood = -3251.2553
```

Rating scale model   Number of obs   =   514
Log likelihood = -3251.2553

	Coef.	Std. Err.	z	P>|z|	[95% Conf. Interval]	
Discrim	1.248191	.0663571	18.81	0.000	1.118133	1.378248
calm						
Diff						
2 vs 1	-.8072838	.0808518	-9.98	0.000	-.9657504	-.6488171
3 vs 2	.5034159	.0782908	6.43	0.000	.3499688	.6568631
4 vs 3	1.373037	.0974431	14.09	0.000	1.182052	1.564022
5 vs 4	2.622023	.1591598	16.47	0.000	2.310076	2.933971
tense						
Diff						
2 vs 1	-1.018429	.0841955	-12.10	0.000	-1.183449	-.853409
3 vs 2	.2922705	.0764788	3.82	0.000	.1423747	.4421662
4 vs 3	1.161892	.0926895	12.54	0.000	.9802236	1.34356
5 vs 4	2.410878	.1535633	15.70	0.000	2.109899	2.711856
regretful						
Diff						
2 vs 1	-.6935233	.0793644	-8.74	0.000	-.8490746	-.537972
3 vs 2	.6171764	.0797475	7.74	0.000	.4608742	.7734786
4 vs 3	1.486798	.1002515	14.83	0.000	1.290308	1.683287
5 vs 4	2.735784	.1622153	16.87	0.000	2.417847	3.05372

```
atease
         Diff
     2 vs 1    -1.034104    .0844713    -12.24    0.000    -1.199665     -.868543
     3 vs 2     .2765959    .0763917      3.62    0.000     .1268709     .4263209
     4 vs 3     1.146217    .0923628     12.41    0.000     .9651894     1.327245
     5 vs 4     2.395203    .1531507     15.64    0.000     2.095033     2.695373

nervous
         Diff
     2 vs 1    -.9020018    .0822632    -10.96    0.000    -1.063235    -.7407689
     3 vs 2     .4086979    .0773321      5.28    0.000     .2571297     .5602661
     4 vs 3     1.278319    .0952326     13.42    0.000     1.091667     1.464972
     5 vs 4     2.527305     .156639     16.13    0.000     2.220298     2.834312
```

In this output, in addition to the model parameter estimates interpretation as in the PCM (see section 11.3), we notice a main characteristic feature of the RSM by observing that, say, the differences among the (estimated) difficulty parameters for the consecutive categories of the "nervous" item are the same as their corresponding ones in any other item. For instance, the discrepancy in the first two difficulty parameter estimates for the "at ease" item is $(-1.03) - 0.28 = -1.31$. This discrepancy is the same as the difference between the first two difficulty parameter estimates for the "nervous" item, which is $(-0.90) - 0.41 = -1.31$. (As earlier in this chapter, all estimates and differences are rounded off to the second digit after the decimal point.)

With this property in mind, we also note from (11.4) that equivalently the cross-item differences in the difficulty parameters are the same across response categories. For example, the difference for the first-listed difficulty parameters of the items "at ease" and "nervous", which is $(-1.03) - (-0.90) = -0.13$, is the same as that difference at each following difficulty parameter across these 2 items. In particular, for the second listed, for the third listed, and for the fourth listed difficulty parameters, the difference in these parameters across the two items considered is -0.13, as can be readily seen from the last presented Stata output; see its panels for the difficulty parameters of the "at ease" and "nervous" items above. Because of this RSM feature, we present below a graph with the CRFs for the first response category, which we obtain with the next command (recall that the lowest response option is denoted "1" in the analyzed dataset, pertaining to the answer "not at all applicable to me"):

```
. irtgraph icc 1.calm 1.tense 1.regretful 1.atease 1.nervous
```

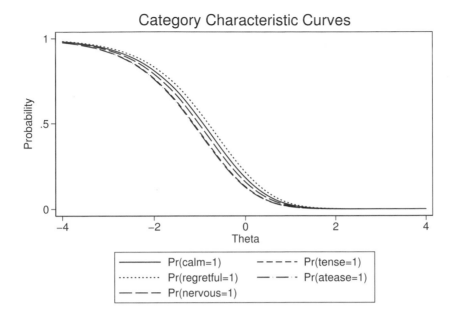

Returning to the last presented Stata output, we notice that similarly to those of the PCM, the difficulty parameters for increasing categories show in the RSM an increasing trend when we compare their confidence intervals within an item. As with the PCM, we note that the difficulty parameters in the RSM represent points on the underlying θ-scale where two adjacent categories are equally probable. For instance, for the "nervous" item, the last difficulty parameter estimate, 2.53, is the point on the θ-scale where a person would have the same probability of choosing "often applicable to me" or "always applicable to me" on this item. (See earlier section in this chapter for the specific meaning of the response options.) To see this feature highlighted on the CRF graph, we use the next command (plot following it):

```
. irtgraph icc nervous, xlabel(-4 2.53 4)
```

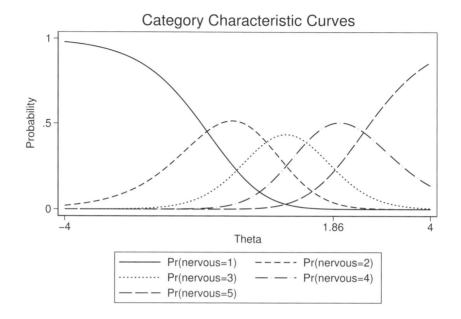

It is worthwhile noting here that because the RSM is more parsimonious than the PCM, "on average" the standard errors of the difficulty parameters in the former will tend to be smaller and therefore their confidence intervals narrower. (For instance, the standard error for the first difficulty parameter of the "nervous" item is 0.110 in the PCM but 0.82 in the RSM.)

In conclusion, we point out that the RSM is applicable for the used anxiety dataset because i) all items in it have the same number of response categories and ii) all of these categories are assumed to have the same meaning across items (see above in this chapter for pertinent details).

11.5   The generalized partial credit model

The previously discussed PCM and RSM are essentially extended Rasch models for the case of polytomous items, as indicated earlier in the chapter. Hence, their common feature is the assumption of equal discrimination parameters across items in a given set or measuring instrument of concern. This assumption is relaxed in another widely used polytomous IRT model, the GPCM (frequently referred to as "GPCM" in the literature, using its abbreviation). The GPCM was proposed by Muraki (1992) and like the PCM, the GPCM is appropriate for modeling responses on items whose successful accomplishment requires correct completion of at least two "elementary" tasks. The GPCM generalizes the PCM by permitting each item to have its own overall discrimination parameter.

That is, the GPCM can be thought of as a PCM in which the item invariance assumption with respect to the discrimination parameter is relaxed.

The GPCM is defined as follows, in terms of probability of response in the qth option for a given item (say, jth) as a function of the underlying unobserved trait θ,

$$P_{jq}(\theta) = \exp\left\{\sum_{s=0}^{q} a_j\left(\theta - b_{js}\right)\right\} \Big/ \left[1 + \sum_{h=1}^{r} \exp\left\{\sum_{s=1}^{h} a_j\left(\theta - b_{js}\right)\right\}\right]$$

where a_j and b_{jq} are item discrimination and category difficulty parameters, respectively ($j = 1, \ldots, k, q = 0, 1, \ldots, r$; see earlier sections on the NRM and PCM for model definition equations and parameter interpretation, including the case $q = 0$ for the PCM).

To illustrate the GPCM empirically, we return once again to the anxiety data analyzed earlier in this chapter and use the following Stata command:

```
. irt gpcm calm-nervous

Fitting fixed-effects model:

Iteration 0:   log likelihood = -3599.4342
Iteration 1:   log likelihood = -3599.4342

Fitting full model:

Iteration 0:   log likelihood = -3378.8602
Iteration 1:   log likelihood = -3295.4418
Iteration 2:   log likelihood = -3208.6593
Iteration 3:   log likelihood = -3187.3069
Iteration 4:   log likelihood = -3186.5826
Iteration 5:   log likelihood = -3186.5784
Iteration 6:   log likelihood = -3186.5784

Generalized partial credit model              Number of obs    =        514
Log likelihood = -3186.5784
```

	Coef.	Std. Err.	z	P>|z|	[95% Conf. Interval]	
calm						
Discrim	1.936805	.2269569	8.53	0.000	1.491978	2.381633
Diff						
2 vs 1	-.9020124	.092316	-9.77	0.000	-1.082948	-.7210764
3 vs 2	.3721059	.0816284	4.56	0.000	.2121172	.5320945
4 vs 3	1.507602	.126089	11.96	0.000	1.260472	1.754732
5 vs 4	2.518992	.239158	10.53	0.000	2.050251	2.987733
tense						
Discrim	1.533478	.1774046	8.64	0.000	1.185771	1.881184
Diff						
2 vs 1	-1.012972	.1087208	-9.32	0.000	-1.226061	-.799883
3 vs 2	.3051627	.0965118	3.16	0.002	.116003	.4943224
4 vs 3	1.041503	.1182395	8.81	0.000	.8097582	1.273248
5 vs 4	2.40014	.2269986	10.57	0.000	1.955231	2.845049

regretful						
Discrim	.7090085	.0840776	8.43	0.000	.5442194	.8737975
Diff						
2 vs 1	-.4709243	.1655744	-2.84	0.004	-.7954441	-.1464045
3 vs 2	1.024336	.2099109	4.88	0.000	.6129185	1.435754
4 vs 3	1.284149	.2476905	5.18	0.000	.7986844	1.769613
5 vs 4	2.792374	.4210748	6.63	0.000	1.967082	3.617665
atease						
Discrim	2.225673	.2736658	8.13	0.000	1.689298	2.762048
Diff						
2 vs 1	-1.151018	.097018	-11.86	0.000	-1.34117	-.9608666
3 vs 2	.1408705	.0736209	1.91	0.056	-.0034238	.2851648
4 vs 3	1.240779	.1027848	12.07	0.000	1.039324	1.442233
5 vs 4	2.604826	.2288308	11.38	0.000	2.156326	3.053326
nervous						
Discrim	.8505312	.0972803	8.74	0.000	.6598653	1.041197
Diff						
2 vs 1	-.648044	.1495251	-4.33	0.000	-.9411078	-.3549801
3 vs 2	.5609441	.1620685	3.46	0.001	.2432956	.8785926
4 vs 3	1.094405	.1928782	5.67	0.000	.7163711	1.47244
5 vs 4	2.514483	.3259446	7.71	0.000	1.875644	3.153323

From this output, we notice easily that the GPCM is associated with an estimated discrimination parameter for each item, unlike the PCM, of which the former model is a generalization in this respect. In particular, the "at ease" item is associated with highest discrimination power in the analyzed dataset (see below). To highlight the difference between the GPCM and the more restrictive PCM, we plot the CRFs for two items under the GPCM, using the `irtgraph icc` Stata command followed by the name of the item in question. For instance, for the "calm" and "tense" items, the resulting CRF graphs are as follows (in this order):

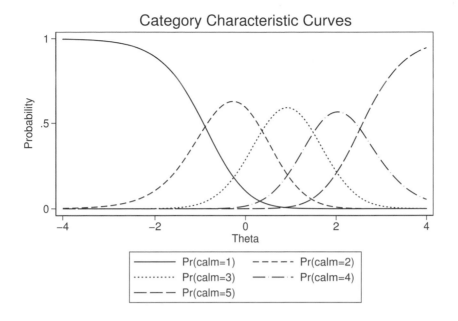

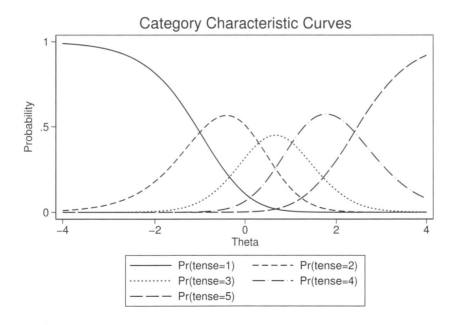

The difference in the item discrimination parameters, which as indicated previously is a characteristic feature of the GPCM, leads to discrepancies across items in the gradient of increase in a subsequent category's CRF immediately after passing the point on the θ-scale where the latter curve crosses that of the preceding response option. This property of the CCCs, resulting from the differing item discrimination parameters, is also seen when plotting the lowest category CCCs simultaneously for all items, using this command (graph immediately following command):

```
. irtgraph icc 1.calm 1.tense 1.regretful 1.atease 1.nervous
```

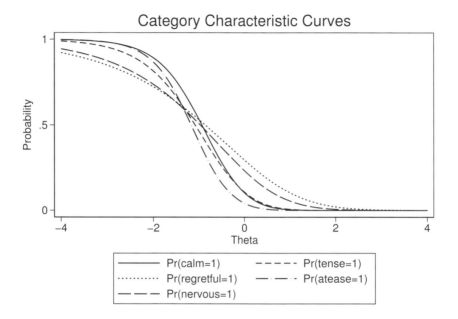

Based on the last graph, an important observation can be made. Specifically, the drop in probability for the lowest category, which is associated with a given increase in the underlying trait around the midprobability point on the θ-scale, is most precipitous in the "at ease" item. The reason is that its (estimated) discrimination parameter is the highest across items, as we can see from the above results of fitting it to the analyzed anxiety data. (Note that if one compares this category with the next higher, in the GPCM, the discrimination parameter acts in the same way as it would in a dichotomous item with only these two response options; thereby, the currently considered category of relevance for the above graph would be denoted "0" and the next higher "1"; see also chapters 1 and 2.) In fact, as seen from the earlier presented output for the GPCM, the 95% confidence interval for the discrimination parameter of the "at ease" item is completely above that interval of the "regretful" and "nervous" items. Hence, as expected, we notice from the last graph the latter two items being associated with the flattest CCCs for the lowest category and, in particular, with CCCs markedly flatter than the CCC for the "at ease" item.

The GPCM discussed in this section follows the same idea of presenting in its defining equation the probability for response category choice as a function of the underlying trait, just as the NRM, PCM, and RSM discussed earlier in the chapter did. This direct definition feature for that probability unifies these four models, so they may be viewed as "direct" probability models (for example, Thissen and Steinberg [1986]). An alternative approach to defining this probability in an indirect way underlies the next considered model that is rather popular in empirical settings with polytomous items in behavioral, educational, and social research.

11.6 The graded response model

A main approach to polytomous IRT models that is distinct from all preceding models goes back to the highly influential work by Samejima (1969). This is the graded response model (usually abbreviated to GRM in the literature). In part due to its wide applicability, for almost half a century now, the GRM has been attracting much interest among methodologists and substantive researchers.

Like the PCM, GPCM, and RSM, the GRM stipulates that for a considered item, the response options are ordered rather than nominal; that is, they fall along an ordinal "scale" of measurement. The field of application of the GRM is perhaps best exemplified by a reference to the highly popular Likert rating scales (Likert items) in the social sciences and well beyond them.

The aim of the GRM, like that of the earlier considered polytomous models, is to use the additional information about the individual trait or ability levels. This information is contained in the knowledge of one's particular answer to a polytomous item and is richer than the information consisting of only knowing whether the selected option by him or her is "correct" or "false", as in the binary or binary scored item case. A main feature of the GRM is that it capitalizes on this additional information. In doing so, the GRM allows one to obtain better—that is, more precise—estimates of the individual ability levels as well as of item parameters.

The GRM can be formally defined with the following expression for the probability of selecting the qth option on the jth item with r ordered categories, in a given set of items of interest ($q = 1, \ldots, r - 1$; $j = 1, \ldots, k$; compare Samejima [2016]):

$$P_{j,q}(\theta) = P_{j,q}^+(\theta) - P_{j,q+1}^+(\theta) \tag{11.5}$$

The two probabilities appearing in the right-hand side of (11.5) differ from the types of response probabilities used thus far in the book. Specifically, $P_{j,q}^+(\theta)$ is the probability of selecting the qth or higher option on the item in question. Similarly, $P_{j,q+1}^+(\theta)$ is the probability of selecting the $(q + 1)$th or higher option on it. (This interpretation is analogous to that of their counterpart probabilities in the ordinal logistic regression setting; see, for example, Agresti [2013].) These two probabilities are assumed in the GRM to be the following ratios:

$$P_{j,q}^+(\theta) = \frac{\exp\{a_j(\theta - b_{j,q})\}}{1 + \exp\{a_j(\theta - b_{j,q})\}} \qquad (11.6)$$

In (11.6), $b_{j,q}$ is a difficulty level parameter for the qth category ($q = 2,\ldots,r; j = 1,\ldots,k$). This parameter is interpretable as a "cutpoint" on the θ-scale that identifies a pertinent boundary between consecutive ordered response categories. Specifically, $b_{j,q}$ represents a point on that scale, which has the following property: if a person possesses ability level equal to that point, that is, $b_{j,q}$, then he or she has a probability of 0.5 to respond to a category q or higher (see also chapter 1). In addition, a_j is a discrimination parameter, which along with $b_{j,q}$ is associated with the jth item. (As mentioned earlier in the chapter, it is assumed that the probability for a response $1, 2, \ldots, r-1$ or r is 1, that is, the probability of a response in the first category here or higher is 1; similarly, the probability of a response higher than r is 0.)

The GRM does not require the same number of categories across all items, as the RSM does (see section 11.5). However, our discussion of the GRM in this section formally assumes for simplicity and convenience the same number of categories, because this does not limit the generality of the developments in this section. As alluded to earlier, the GRM can be considered an analog of the ordinal logistic regression model (for example, Agresti [2013]) for the case of an unobserved predictor. This permits the above interpretation of the GRM parameter estimates in terms of the dichotomy of an item response up to a given option versus an answer being either that option or an option above it.

The right-hand side of (11.6) bears marked resemblance to that of the defining equation of the 2PL model (see chapter 5). Specifically, with only $r = 2$ response options, the GRM is equivalent to the 2PL model. More concretely, with two categories for an item under consideration, (11.5) and (11.6) yield the definition of the 2PL model [see also discussion after (11.6), and recall that we then need only the ICC for one of the response categories, that is, the "higher", because that for the other is its complement to 1]. With this in mind, the GRM can also be seen as an extension of the 2PL model to the case of polytomous ordinal items. Further, we stress that (11.5) involves the subtraction of two probabilities of the type presented in (11.6). Thus, the GRM can also be characterized as a "difference model", or an "indirect" IRT model (for example, Thissen and Steinberg [1986]).

As an example to illustrate the GRM, let us consider again the anxiety study data that we used earlier in the chapter. To fit the GRM to the five anxiety items' data, we use the following Stata command, which is followed by the output produced thereby:

```
. irt grm calm-nervous

Fitting fixed-effects model:

Iteration 0:   log likelihood = -3608.5847
Iteration 1:   log likelihood = -3599.5937
Iteration 2:   log likelihood = -3599.4343
Iteration 3:   log likelihood = -3599.4342

Fitting full model:

Iteration 0:   log likelihood = -3357.6872
Iteration 1:   log likelihood = -3183.6327
Iteration 2:   log likelihood = -3175.5036
Iteration 3:   log likelihood = -3174.8601
Iteration 4:   log likelihood = -3174.8578
Iteration 5:   log likelihood = -3174.8578
```

Graded response model   Number of obs   =   514
Log likelihood = -3174.8578

	Coef.	Std. Err.	z	P>|z|	[95% Conf. Interval]	
calm						
Discrim	2.595019	.2426888	10.69	0.000	2.119357	3.07068
Diff						
>=2	-.9167881	.0823798	-11.13	0.000	-1.07825	-.7553267
>=3	.3278215	.0669774	4.89	0.000	.1965482	.4590949
>=4	1.503184	.1060512	14.17	0.000	1.295327	1.71104
=5	2.654593	.1984592	13.38	0.000	2.26562	3.043566
tense						
Discrim	2.132803	.1902481	11.21	0.000	1.759923	2.505682
Diff						
>=2	-1.053258	.0940871	-11.19	0.000	-1.237666	-.868851
>=3	.2016509	.0701386	2.88	0.004	.0641819	.33912
>=4	1.171864	.0975893	12.01	0.000	.9805928	1.363136
=5	2.486355	.1876887	13.25	0.000	2.118492	2.854218
regretful						
Discrim	1.229264	.125503	9.79	0.000	.9832824	1.475245
Diff						
>=2	-.8097067	.1150251	-7.04	0.000	-1.035152	-.5842616
>=3	.6833971	.1070904	6.38	0.000	.4735038	.8932904
>=4	1.748906	.174671	10.01	0.000	1.406557	2.091255
=5	3.242628	.3256212	9.96	0.000	2.604423	3.880834
atease						
Discrim	2.843415	.2754609	10.32	0.000	2.303521	3.383308
Diff						
>=2	-1.144514	.0878638	-13.03	0.000	-1.316724	-.972304
>=3	.1213664	.0634824	1.91	0.056	-.0030569	.2457896
>=4	1.264687	.0917707	13.78	0.000	1.084819	1.444554
=5	2.618167	.1956098	13.38	0.000	2.234779	3.001555
nervous						
Discrim	1.432981	.1355271	10.57	0.000	1.167353	1.698609
Diff						
>=2	-.9297679	.1094188	-8.50	0.000	-1.144225	-.7153111
>=3	.3817735	.0876504	4.36	0.000	.2099819	.553565
>=4	1.402345	.132558	10.58	0.000	1.142536	1.662154
=5	2.774817	.2488187	11.15	0.000	2.287142	3.262493

We notice from this output that for each item, the GRM is associated with a discrimination parameter and, in the present example, four difficulty parameters. Thereby, those of the latter four parameters that pertain to higher categories are larger than the ones for lower categories, as can be expected because, as mentioned, the GRM is a counterpart to a corresponding ordered logistic regression model (see also chapters 4 and 5). In addition, the difficulty parameters are increasing within an item following a numerical pattern that is fairly similar across the five items. The interpretation of each of the four difficulty parameters per item is, as mentioned above, as that position on the underlying θ scale, at which a person with a trait level equal to it has a probability of 1/2 for responding in the pertinent category (associated with that parameter) or higher (see also section 2.3 in chapter 2). For example, looking at the estimate of 0.38 (rounded off) for the second-listed category of the "nervous" item in the last presented Stata output, we can say that a person with that trait level, that is, of 0.38, has a probability of 0.5 for answering i) with response category 1 or 2 versus ii) with response category 3 or higher (that is, with category 3, 4, or 5). Similarly, someone with a trait level of -0.93 has the same probability to answer 1 as to answer 2 or higher. Also, a person with a trait level of 1.40 has the same chance of answering 4 or higher as for answering 1, 2, or 3. Lastly, someone located at 2.77 on the θ-scale has a probability of 0.5 to answer 5 and the same probability to answer 1, 2, 3, or 4 on this item.

To highlight further this interpretation, we plot next the so-called boundary characteristic curves (BCCs) for the **nervous** item. The BCCs can be plotted using the Stata command given next, which is reminiscent of the way we obtained the ICCs for dichotomous items in earlier chapters of the book. The reason is that the BCC is the corresponding ICC when considering i) the pertinent response in a given category or higher (if it is not the highest already); versus ii) a lower category (if the former category is not the lowest). These response categories are correspondingly indicated in the legend of the resulting plot produced by the following command:

```
. irtgraph icc nervous, blocation
```

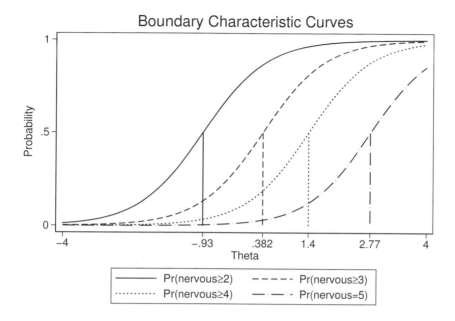

We see from this plot that the midprobability points (projections) on the θ-scale equal the estimated difficulty level parameters for the item in question (see also last panel of above presented output for the fit model and our discussion preceding this graph). This is because these BCCs, as mentioned, are the ICCs if one were to consider the item responses as dichotomized into successive pairs of categories for answering with at least (the choice denoted) q and answering with an option less than q $(q = 2, \ldots, r)$. We also note that these BCCs are parallel in their central part, which is because there is a single discrimination parameter for any given item (and only the BCC for the "nervous" item are presented in the last plot). Hence, the same discrimination parameter would be applicable for all of those successive dichotomizations (associated with the displayed BCCs for an item).

The different discrimination parameters across items lead to graphs with BCCs possessing different steepness in their central part. We notice this readily by comparing the last graph with that, say, for the "at ease" item (obtained with the same last Stata command where atease is used at its very end in lieu of nervous):

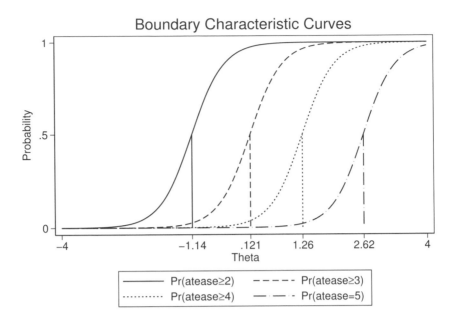

We observe that the gradient at the midprobability point of any of the BCCs for the "at ease" item is markedly higher than the gradient at the same point for any of the BCCs of the "nervous" item. This notable difference is a result of the considerable difference in their discrimination parameter estimates, as seen from the GRM output presented above. In particular, we notice that the 95% confidence interval for the discrimination parameter of the "at ease" item is entirely above that interval for the "nervous" item. This leads also to the marked difference in the steepness (in the central part) of their associated BCCs in the last two presented graphs.

Because the GRM is as mentioned a "difference model" (Thissen and Steinberg 1986), in addition to the above BCCs, we can furnish item-specific CCCs. The CCCs, as mentioned earlier in this section, represent the probability of a response in a given category as a function of the underlying trait or ability θ. These individual response category probabilities are readily obtained for each item in terms of appropriate differences, stated formally in (11.5), as functions of θ. The CCCs can be plotted in the following way, for instance, for the **nervous** item (graph follows immediately after command):

```
. irtgraph icc nervous, xlabel(-4 -3 -2 -1 0 1 2 3 4, grid)
```

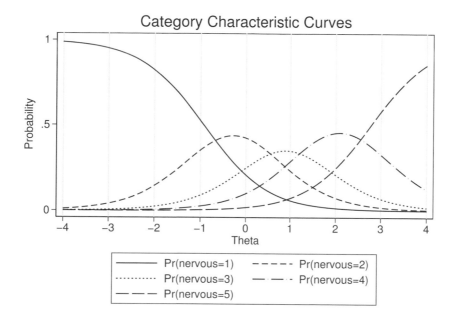

We see from this graph that respondents with a latent trait level below approximately −0.7 are most likely (have highest probability) to respond in the lowest category (denoted "1" in the dataset), that is, "not at all applicable to me". Similarly, respondents with a trait level between approximately −0.7 and 0.5 are most likely to respond with the second category, "rarely applicable to me". Further, those with anxiety levels between 0.5 and 1.2 (approximately) are most likely to select the third category, "sometimes applicable to me". Also, persons with anxiety levels between 1.2 and 2.5 (approximately) have as the most probable response the fourth category, "often applicable to me". Lastly, respondents with trait levels above approximately 2.5 are most likely to respond with the fifth category, "always applicable to me". With these observations in mind, we can view the points where the CCCs cross for adjacent categories as the locations on the underlying latent dimension at which in a sense "transitions" from one category to the next occur. Specifically, immediately after these "transition" points, the probability of response in the higher category is for the first time larger than that for the lower category (for a given pair of adjacent response categories in an item of interest).

If viewing the 5 studied items as building a scale measuring anxiety, with a total score ranging between 5 and 25, we can also examine what kind of scale scores one can expect from persons with different levels of anxiety. To this end, we can inspect the test characteristic curve (TCC) for this anxiety scale in the following way (resulting graph presented after Stata command):

. irtgraph tcc, thetalines(-1.96 0 1.96)

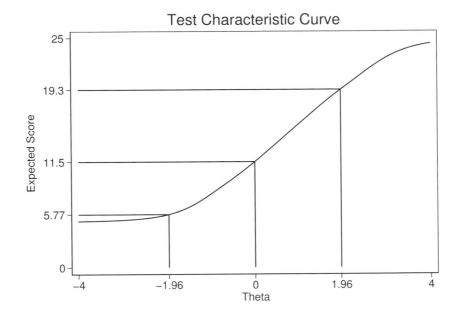

From this TCC plot, we see that 1 out of 20 respondents in the long run can be expected to score on this overall scale either 1 through 5, or at least 20. The remaining subjects then can be expected to score between 6 and 19. We would also expect above-average persons on anxiety to score 12 or above, while below-average subjects would be expected to score no more than 11.

Similarly, if a researcher is interested in obtaining the expected total scores that persons within certain ranges on the underlying latent dimension θ could receive (under the assumption of normality of the studied population on it), he or she could proceed in the following way. The researcher could request a plot of the TCC with a superimposed grid of "theta lines" at consecutive standard deviations below or above the mean on that dimension. This is accomplished with the following Stata command, for instance, for the overall interval of 6 standard deviations around the studied ability mean of 0 (resulting graph following it):

```
. irtgraph tcc, thetalines(-3/3)
```

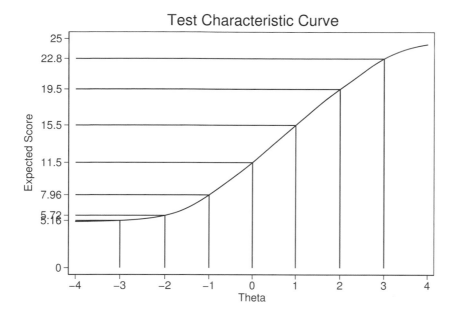

As seen from the last plot, persons that are, say, between 1 and 3 standard deviations above the mean on anxiety can be expected to score between 16 and 22 on the scale consisting of the 5 considered items.

While these are interesting and potentially insightful interpretations for the substantive researcher in particular, returning for a moment to the above output with the GRM estimation results, we note that it is associated with an impressive number of parameters. Indeed, counting the parameters across items, we have actually estimated a total of $5 + 5 \times 4 = 25$ model parameters in the currently considered anxiety example. This number makes the GRM obviously far less parsimonious than, say, the RSM, as well as notably less parsimonious than the PCM (see preceding discussion in this chapter). At the same time, we also realize that the quality of a model depends critically on its fit to the analyzed data. Therefore, a simple question that arises now is this: while the GRM may be much less parsimonious than, say, the RSM, is the fit of the former sufficiently better than that of the latter to render the GRM still preferable to the RSM?

More generally, having considered five polytomous IRT models in the chapter, we must next address how to choose among them all. One important criterion to use is of course substantive and study design-related considerations. For instance, if it is clear that the items in a given set or measuring instrument are nominal and not ordinal, it would generally make little if any sense to fit to them a model other than the NRM. Conversely, if the items in the considered set are ordinal (that is, with ordered categories), the NRM can be ruled out as a candidate model. Further, if the items have a different number of ordered categories, it would not be sensible to consider fitting the RSM. In addition to these three simple study design-related criteria, we recommend using popular model-selection criteria that we have already used previously in the book and are available to use also for polytomous IRT model choice. In the next section, we address the above query—whether we can place more trust in the GRM or in any of the previously fit models in this chapter—after conducting information criteria-based comparison of all five discussed models.

11.7   Comparison and selection of polytomous item response models

To become involved in model selection from the above set of five models that we have considered for the analyzed anxiety dataset, as indicated earlier (see chapter 6), we can use the widely used Akaike information criterion (AIC) and Bayesian information criterion (BIC) indices. As we mentioned and illustrated in chapter 6, they are obtained using the Stata command `estat ic` after each of the models is fit. To make the discussion in this section more complete, we list the needed Stata commands to accomplish the AIC and BIC evaluation next:

```
. quietly irt nrm calm-nervous
. estat ic
  (output omitted )
. quietly irt pcm calm-nervous
. estat ic
  (output omitted )
. quietly irt rsm calm-nervous
. estat ic
  (output omitted )
. quietly irt gpcm calm-nervous
. estat ic
  (output omitted )
. quietly irt grm calm-nervous
. estat ic
  (output omitted )
```

In this way, we obtain their following information indices (see table 11.1).[2]

---

2. If one is willing to argue that the available response options for each of the five items considered are ordinal to begin with, rather than nominal, then the NRM can obviously be excluded from the model comparison activity conducted in this section (see also next section).

Table 11.1. Log likelihoods, Akaike information criterion indices, Bayesian information criterion indices, and number of parameters of the nominal response model, partial credit model, rating scale model, generalized partial credit model, and graded response model fit to the anxiety data (t = number of model parameters, n = sample size, Log_like = maximized log likelihood)

	Log_like	t	AIC	n	BIC
NRM	−3179.511	40	6439.022	514	6608.711
PCM	−3217.319	21	6476.638	514	6565.724
RSM	−3251.255	9	6520.511	514	6558.691
GPCM	−3186.578	25	6423.157	514	6529.212
GRM	−3174.858	25	6399.716	514	6505.771

Because both the AIC and BIC indices of the GRM are the smallest, in terms of overall fit and model complexity (see chapter 6 for their definition), we favor the GRM over the NRM, PCM, RSM, and GPCM. With this choice of the GRM for the analyzed anxiety dataset, in particular also when compared with the NRM, one may suggest that the analyzed dataset is consistent with the following suggestion: the five items from the studied scale could be considered ordinal measures of particular aspects of anxiety. However, a final "verdict" on such a statement can only be determined by an expert in the subject-matter domain of anxiety, rather than based exclusively on the above statistical considerations.

Given that we have preferred the GRM based on statistical criteria, it would be of interest next to look at the individual IIFs as well as the overall test information function and TCC associated with it. (Use the next command for each item in turn, immediately after fitting the GRM to the anxiety dataset; IIFs follow this command.)

. irtgraph iif calm

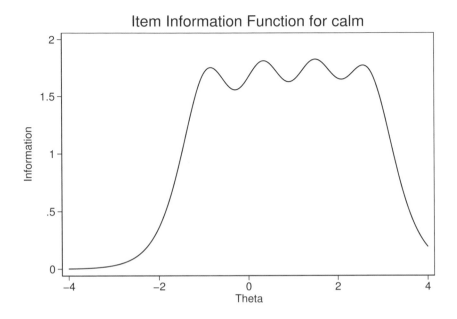

. irtgraph iif tense

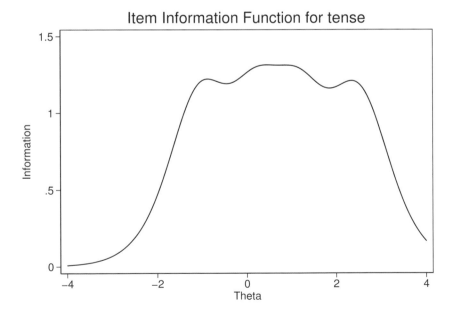

. irtgraph iif regretful

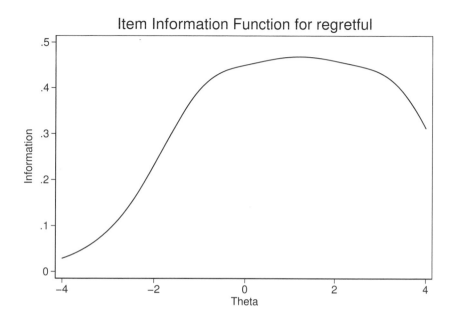

. irtgraph iif atease

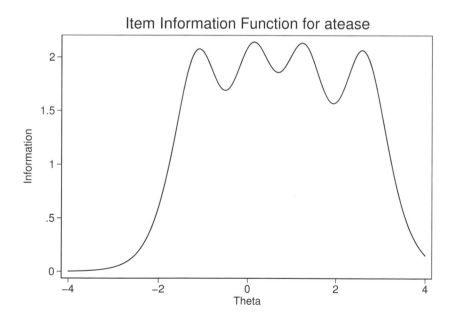

```
. irtgraph iif nervous
```

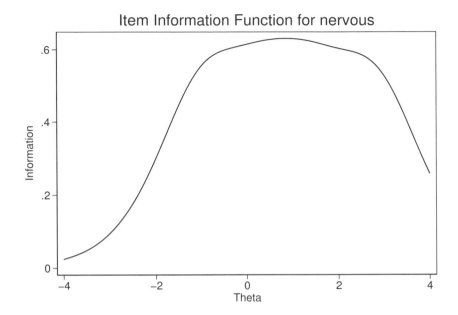

We observe that all five items provide relatively high amounts of information for average and (slightly) above-average levels of anxiety evaluated by them. In addition, we note that the IIF under the GRM need not be unimodal or symmetric. The reason is that the IIF is made up of contributions from the individual categories, and these may differ substantially within a given item.

When we consider the analyzed set of five items as an overall scale of anxiety, say, it would be of interest to examine next the associated test information function, which is obtained in the following way:

. irtgraph tif

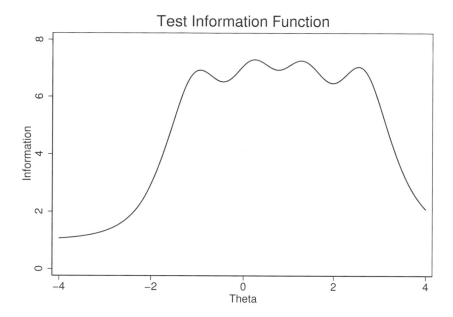

In this figure, we observe that the scale similarly offers the highest information for approximately the same range of individual trait, construct, or latent levels as the individual items do, as could be expected based on the above IIFs, which are highest over approximately the same range on the underlying continuum. Further, the last figure suggests that the scale consisting of the five items under consideration could be used as an instrument evaluating with highest precision the majority of the population from which the examined persons came, excluding the subpopulation of low- and of very high-anxiety persons.

For empirical work with possibly more pragmatic aims, one could consult the TCC of this five-item anxiety scale as well, which we have obtained and discussed in section 11.6 for the GRM, that we have selected in this section as the preferred model for the analyzed dataset.

11.8   Hybrid models

Up to this point in the book, we have considered IRT models where all items were assumed to possess ICCs belonging to the same functional class, with possibly differing values of their associated parameters. For instance, in the preceding chapters (1 through 10), all items were binary or binary scored. Similarly, so far in this chapter, all items were assumed to be either nominal or all ordinal. Moreover, when fitting an IRT model, we assumed all items were following the same model. For instance, earlier in the book,

q2

Discrim						
2 vs 1	-.1940619	.2058634	-0.94	0.346	-.5975466	.2094229
3 vs 1	.5554921	.2001948	2.77	0.006	.1631175	.9478668
4 vs 1	-.0926764	.1957148	-0.47	0.636	-.4762703	.2909175
Diff						
2 vs 1	5.986221	6.587642	0.91	0.364	-6.925319	18.89776
3 vs 1	-3.529379	1.256797	-2.81	0.005	-5.992655	-1.066102
4 vs 1	17.80328	37.97814	0.47	0.639	-56.63251	92.23907

q3

Discrim						
2 vs 1	-.2375184	.2286105	-1.04	0.299	-.6855868	.2105499
3 vs 1	.7013507	.2196845	3.19	0.001	.270777	1.131924
4 vs 1	1.274853	.2460029	5.18	0.000	.7926961	1.75701
Diff						
2 vs 1	1.12829	1.74653	0.65	0.518	-2.294845	4.551425
3 vs 1	-1.824877	.4596287	-3.97	0.000	-2.725732	-.9240208
4 vs 1	-1.131442	.1882568	-6.01	0.000	-1.500419	-.7624658

pcm

q4

Discrim	.3300807	.0526185	6.27	0.000	.2269504	.4332111
Diff						
2 vs 1	-2.056822	.322035	-6.39	0.000	-2.687999	-1.425645
3 vs 2	-.2976236	.1923535	-1.55	0.122	-.6746295	.0793823
4 vs 3	.8472731	.2048051	4.14	0.000	.4458626	1.248684

In the last output with model results, we first note the separation of the NRM from the PCM portion, which is important to keep in mind when interpreting the model parameter estimates. We also notice the increasing discrimination parameters associated with the third nominal item, while the difficulty parameters of the other two nominal items do not show any particular trend (which is not unexpected given their nominal nature). Further, for the third item, the discrimination power of the category choice 3 versus 1 is notably stronger than that of choice 2 versus 1. This is also suggested by comparing their 95% confidence intervals, noting that the confidence interval of the former discrimination parameter is entirely above that interval for the latter discrimination parameter (see also section 11.3). The marked increase in the difficulty parameters for item 4 is also noticed readily, based on the same argument. This is consistent with an interpretation of increasingly higher levels of the (presumed) physical science ability required for selection of the higher categories on this item (which likely correspond to more complete item responses in that empirical study). This pattern is also observed when inspecting the CRFs of item 4 (graph follows command):

. irtgraph icc q4

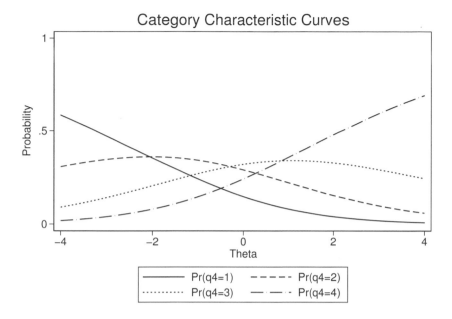

This graph demonstrates as expected the monotonically decreasing nature of the characteristic curve for the lowest category (category 1), as well as the monotonically increasing nature of the characteristic curve for the highest category (category 4). However, the middle two categories do not follow either pattern and remain nearly uniformly high throughout much of the depicted range of the underlying ability. The test information function, presented next, shows that the overall science test, consisting of the 4 considered items, provides the highest precision for evaluation of the individual ability levels of a sizable section of below-average students and, specifically, for those positioned between 2 standard deviations below the studied ability mean and the latter:

```
. irtgraph tif
```

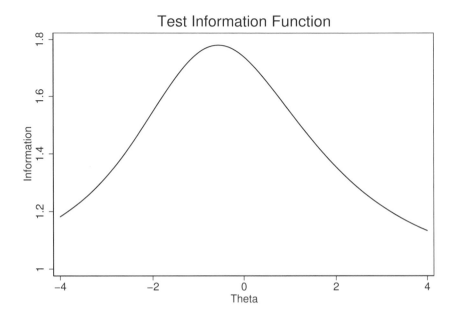

We conclude this section on hybrid models by stressing that the approach discussed in it has very wide applicability in empirical research. As illustrated, Stata offers a readily used means to fit hybrid IRT models.

11.9   The three-parameter logistic model revisited

In chapter 5, we defined the three-parameter logistic (3PL) model as an extension (generalization) of the 2PL model. This extension included a pseudoguessing parameter accounting for a possible chance-based correct response on some or all of the items, as well as the fact that some examined persons with potentially rather low ability level may provide the correct answer as a result of chance-driven processes. As will be recalled, we indicated that in general, the 3PL model can be associated with numerical difficulties when fit to empirical data. We also considered a parsimonious 3PL model, in which the pseudoguessing parameter was assumed invariant across all items of concern. The more general version of that model, where each item (with possible guessing) has its own pseudoguessing parameter, can be fit in a more straightforward way to data using Bayesian statistics (for example, Cai and Thissen [2015]). However, an application of the hybrid IRT modeling approach from the last section may also be possible for the same aim, as we discuss in this section.

As alluded to above, when a 3PL model is considered in an empirical setting, one need not have the expectation that each item in a set or instrument in question needs to have a pseudoguessing parameter associated with it. Rather, some of the items may have such a parameter, but others may not be associated with any guessing because the latter may not be possible given the nature of those items. For instance, free-response items, those on a test following highly effective instruction, or items with no inherently "correct" or "incorrect" responses cannot be generally thought of as being associated with possible guessing.

With this in mind, we see that in an instrument of interest, one may have several items that could follow, say, the 2PL model or the 1PL model, as items not being associated with possible guessing, while the remaining items may follow a 3PL model. An important question then may well be which particular items may be following the latter model (referred to as group 1 items, say), and which a 2PL or 1PL model (referred to as group 2 items). Once this question is answered, one may fit a hybrid IRT model to the entire set of items or measuring instrument of concern. In that model, the first group of items follows a 3PL model, while the second item group may be complying with a 2PL model or a 1PL model.

To respond to the question of locating the items with possible guessing and those without, one may proceed in the following way. Start with the general 3PL model allowing item-specific pseudoguessing parameters to be included but requesting only a limited number of initial iterations for the fixed-effects model aimed at estimating the item parameters (see, for example, chapter 7). In the resulting solution, the items with nearly zero pseudoguessing parameters are the main candidates for membership in the above group 2 of items with no such parameters. Once having located them, which may be seen as completing step 1 of the present process of fitting a general 3PL model, one uses in step 2 a hybrid IRT model. In it, as indicated above, the last-mentioned items are fit using a 2PL model (or even a 1PL model subsequently, if need be), whereas the remaining items are fit using a 3PL model.

To illustrate this approach to 3PL model fitting, we use adapted data from a mathematics ability study (compare De Boeck and Wilson [2004]; see also *Stata Item Response Theory Reference Manual* [2017a]). The data are found in the file `3pl_s.dta`. Upon accessing it, we carry out the above step 1 with Stata as follows (output presented immediately after command):

```
. use http://www.stata-press.com/data/cirtms/3pl_s.dta

. irt 3pl q1-q9, sepguessing startvalues(iterate(10))

Fitting fixed-effects model:

Iteration 0:   log likelihood = -6228.5704
Iteration 1:   log likelihood = -4999.8728  (not concave)
Iteration 2:   log likelihood = -4973.8918  (not concave)
Iteration 3:   log likelihood = -4973.7597  (not concave)
Iteration 4:   log likelihood = -4973.5365  (not concave)
Iteration 5:   log likelihood =  -4973.536  (not concave)
Iteration 6:   log likelihood = -4973.5359  (not concave)
Iteration 7:   log likelihood = -4973.5359  (not concave)
Iteration 8:   log likelihood = -4973.5359  (not concave)
Iteration 9:   log likelihood = -4973.5359  (not concave)
Iteration 10:  log likelihood = -4973.5359  (not concave)
```

```
Fitting full model:
Iteration 0:   log likelihood = -4920.3946  (not concave)
Iteration 1:   log likelihood = -4867.9402  (not concave)
Iteration 2:   log likelihood = -4818.2745  (not concave)
Iteration 3:   log likelihood = -4794.6883  (not concave)
Iteration 4:   log likelihood = -4789.5387  (not concave)
Iteration 5:   log likelihood = -4788.1812  (not concave)
Iteration 6:   log likelihood = -4787.5156  (not concave)
Iteration 7:   log likelihood = -4786.9119  (not concave)
Iteration 8:   log likelihood = -4786.4831  (not concave)
Iteration 9:   log likelihood = -4786.1314  (not concave)
Iteration 10:  log likelihood = -4785.8152  (not concave)
Iteration 11:  log likelihood = -4785.5382  (not concave)
Iteration 12:  log likelihood = -4785.2833
Iteration 13:  log likelihood = -4785.0129
Iteration 14:  log likelihood = -4784.3544
Iteration 15:  log likelihood = -4783.1278  (not concave)
Iteration 16:  log likelihood = -4782.9846
Iteration 17:  log likelihood = -4782.7662
Iteration 18:  log likelihood = -4782.4864
Iteration 19:  log likelihood = -4782.2669
Iteration 20:  log likelihood = -4782.1941
Iteration 21:  log likelihood = -4782.0807
Iteration 22:  log likelihood = -4782.0274
Iteration 23:  log likelihood = -4781.9632  (not concave)
Iteration 24:  log likelihood = -4781.9543  (not concave)
Iteration 25:  log likelihood = -4781.9468
Iteration 26:  log likelihood = -4781.9355
Iteration 27:  log likelihood = -4781.9247
Iteration 28:  log likelihood = -4781.9194
Iteration 29:  log likelihood = -4781.9127  (not concave)
Iteration 30:  log likelihood = -4781.9114
Iteration 31:  log likelihood = -4781.9105
Iteration 32:  log likelihood = -4781.9086  (not concave)
Iteration 33:  log likelihood = -4781.9085
Iteration 34:  log likelihood = -4781.9082
Iteration 35:  log likelihood = -4781.9073
Iteration 36:  log likelihood = -4781.9071
Iteration 37:  log likelihood = -4781.9067
Iteration 38:  log likelihood = -4781.9065
Iteration 39:  log likelihood = -4781.9064
Iteration 40:  log likelihood = -4781.9064
Iteration 41:  log likelihood = -4781.9064
```

Three-parameter logistic model   Number of obs   =   932
Log likelihood = -4781.9064

	Coef.	Std. Err.	z	P>|z|	[95% Conf.	Interval]
q1						
Discrim	2.699574	.9096512	2.97	0.003	.9166908	4.482458
Diff	-.0536632	.1645703	-0.33	0.744	-.3762152	.2688887
Guess	.173373	.088464	1.96	0.050	-.0000132	.3467592
q2						
Discrim	.6571074	.1046524	6.28	0.000	.4519925	.8622223
Diff	-.1424891	.1129715	-1.26	0.207	-.3639092	.078931
Guess	9.22e-06	.0057202	0.00	0.999	-.0112022	.0112206

q3							
Discrim	.9793953	.1510173	6.49	0.000	.6834068	1.275384
Diff	-1.622865	.2056336	-7.89	0.000	-2.025899	-1.21983
Guess	8.60e-11	.00001	0.00	1.000	-.0000196	.0000196

q4							
Discrim	1.621056	.5960178	2.72	0.007	.4528829	2.78923
Diff	.8535213	.1605929	5.31	0.000	.538765	1.168278
Guess	.2204724	.0657128	3.36	0.001	.0916776	.3492671

q5							
Discrim	.9196113	.1392353	6.60	0.000	.6467151	1.192508
Diff	1.590153	.2049441	7.76	0.000	1.18847	1.991836
Guess	2.23e-13	1.92e-07	0.00	1.000	-3.76e-07	3.76e-07

q6							
Discrim	1.84252	.592093	3.11	0.002	.682039	3.003001
Diff	.8756881	.1172217	7.47	0.000	.6459379	1.105438
Guess	.1428262	.0488265	2.93	0.003	.047128	.2385243

q7							
Discrim	1.493066	1.631757	0.92	0.360	-1.705118	4.69125
Diff	2.355444	.7953906	2.96	0.003	.7965073	3.914381
Guess	.2086883	.0533265	3.91	0.000	.1041702	.3132063

q8							
Discrim	1.297009	.2051395	6.32	0.000	.8949434	1.699075
Diff	-1.775578	.196202	-9.05	0.000	-2.160127	-1.391029
Guess	2.77e-13	3.08e-07	0.00	1.000	-6.03e-07	6.03e-07

q9							
Discrim	.6276239	.110157	5.70	0.000	.4117201	.8435277
Diff	-1.497837	.2562611	-5.84	0.000	-2.000099	-.9955743
Guess	6.87e-10	.0000289	0.00	1.000	-.0000566	.0000566

Before we discuss this output, we note that the last used Stata command includes the separate pseudoguessing parameter request as an option (the `sepguessing` option after the comma). Also, that command requests 10 iterations toward the estimation of the item parameters. At that point, the remainder of the estimation process is initiated as usual. This process leads to the solution presented, after some initial numerical difficulties (that could be expected as mentioned earlier with a 3PL model), yet with a final set of uneventful iterations. However, we do not consider the resulting model solution as the one that is sought. As indicated above, we use it merely to develop a reasonable idea, from a statistical viewpoint, as to which of the items may be associated with some guessing and which are unlikely to be so.

Examining the last presented output results, we hypothesize now that items 1, 2, 3, 5, 8, and 9 are not likely to be associated with any guessing. This is because their pseudoguessing parameter estimates are essentially 0. (We note that they do not appear to be significant, but we are not particularly interested in this finding per se because our aim is only to use this solution to "locate" items in the above groups 1 and 2 correspondingly consisting of items possibly with guessing and items possibly with no guessing.) The last listed are thus the items in group 2, and the remaining items (with pseudoguessing parameters that are not close to 0), that is, items 4, 6, and 7, build group 1 in the model fit next.

In step 2, therefore, using the hybrid modeling approach of the preceding section, we simultaneously fit i) a 2PL model to items 1, 2, 3, 5, 8, and 9 and ii) a 3PL model to items 4, 6, and 7, allowing thereby separate pseudoguessing parameters for them. We achieve this hybrid model fit with Stata as follows (see section 11.8 for related command details):

```
. irt hybrid (2pl q2-q3 q5 q8-q9) (3pl q1 q4 q6-q7, sepguessing),
>    startvalues(iterate(10))

Fitting fixed-effects model:

Iteration 0:   log likelihood = -5665.9274
Iteration 1:   log likelihood = -4978.9692  (not concave)
Iteration 2:   log likelihood = -4973.6868  (not concave)
Iteration 3:   log likelihood = -4973.6205  (not concave)
Iteration 4:   log likelihood = -4973.5414  (not concave)
Iteration 5:   log likelihood = -4973.5364  (not concave)
Iteration 6:   log likelihood =  -4973.536  (not concave)
Iteration 7:   log likelihood = -4973.5359  (not concave)
Iteration 8:   log likelihood = -4973.5359  (not concave)
Iteration 9:   log likelihood = -4973.5359  (not concave)
Iteration 10:  log likelihood = -4973.5359  (not concave)

Fitting full model:

Iteration 0:   log likelihood = -4917.9297  (not concave)
Iteration 1:   log likelihood = -4802.8966
Iteration 2:   log likelihood = -4782.9603
Iteration 3:   log likelihood = -4782.0395
Iteration 4:   log likelihood = -4781.9522
Iteration 5:   log likelihood = -4781.9169
Iteration 6:   log likelihood = -4781.9068
Iteration 7:   log likelihood = -4781.9064
```

```
Hybrid IRT model                          Number of obs     =        932
Log likelihood = -4781.9064
```

	Coef.	Std. Err.	z	P>|z|	[95% Conf.	Interval]
2pl						
q2						
Discrim	.6571222	.1045766	6.28	0.000	.4521557	.8620886
Diff	-.1424124	.1113465	-1.28	0.201	-.3606475	.0758228
q3						
Discrim	.9794136	.1510181	6.49	0.000	.6834236	1.275404
Diff	-1.62284	.2056274	-7.89	0.000	-2.025863	-1.219818
q5						
Discrim	.9195794	.1392297	6.60	0.000	.6466942	1.192465
Diff	1.590198	.2049507	7.76	0.000	1.188502	1.991894
q8						
Discrim	1.29698	.2051369	6.32	0.000	.8949191	1.699041
Diff	-1.7756	.1962091	-9.05	0.000	-2.160163	-1.391037
q9						
Discrim	.6276168	.1101568	5.70	0.000	.4117135	.8435201
Diff	-1.497848	.2562661	-5.84	0.000	-2.00012	-.9955756
3pl						
q1						
Discrim	2.700223	.9100688	2.97	0.003	.9165212	4.483925
Diff	-.0535499	.1645439	-0.33	0.745	-.3760501	.2689502
Guess	.1734329	.0884477	1.96	0.050	.0000785	.3467873
q4						
Discrim	1.620864	.5958152	2.72	0.007	.4530875	2.78864
Diff	.8534893	.1606059	5.31	0.000	.5387075	1.168271
Guess	.2204513	.0657125	3.35	0.001	.0916572	.3492455
q6						
Discrim	1.842788	.5920076	3.11	0.002	.6824743	3.003101
Diff	.8757298	.117204	7.47	0.000	.6460141	1.105445
Guess	.1428561	.0488039	2.93	0.003	.0472022	.2385101
q7						
Discrim	1.497106	1.631168	0.92	0.359	-1.699923	4.694136
Diff	2.353742	.7917429	2.97	0.003	.8019546	3.90553
Guess	.2088318	.0529948	3.94	0.000	.104964	.3126997

We stress that while being a more general model than a 2PL model (for all nine items), the last fit is a hybrid IRT model. As such, it has two groups of items—with and without pseudoguessing parameters—that are separated clearly in the output. We also emphasize that the model does not implement a constraint of equal pseudoguessing parameters for the items with (nonzero) parameters. In this respect, the hybrid model is more general than a 3PL model with an invariant pseudoguessing parameter across all nine items (compare chapter 6). The outlined two-step approach in this section therefore represents a relatively straightforward procedure, based on statistical considerations, for fitting more general 3PL models without the need of invoking assumed prior distributions for their pseudoguessing parameters (compare Cai and Thissen [2015]).

We conclude this section by pointing out the following limitation of the modeling procedure discussed in it. Because of its distinct exploratory element, this method cannot be really used for ascertaining (demonstrating) whether there is guessing on a given item in a measuring instrument or an item set under consideration. That is, this approach cannot be safely used for determining whether an item is indeed associated with guessing on part of the examined persons to whom it is administered. Thus, one will have more confidence in the results of its application when they are consistent with those obtained subsequently on another independent sample from the same studied subject population, for example, via use of the same IRT hybrid model reached with this analytic procedure. This confidence may be further enhanced after carrying out model comparison with other rival models of potential interest, fit to the same dataset, that are then not decided for, for instance, using popular information criteria such as the AIC and BIC (for example, section 11.7).

11.10   Chapter conclusion

In this chapter, we covered a fair amount of ground in polytomous IRT as well as hybrid IRT models. The five models discussed initially represent popular item response modeling tools for nominal and for Likert-type items. They provide a solid means for modeling polytomous items by empirical behavioral and social researchers as well as scholars in cognate disciplines. As illustrated in section 11.7, the model that may be arguably considered often appropriate to start with in an empirical setting with ordinal polytomous items could be the GRM. (However, we do not take that example as sufficient or justification for a stronger and more general model choice recommendation.) We then dealt with a hybrid IRT modeling approach. We recommend the latter for settings where some items are associated with ICCs belonging to a functional class distinct from those pertaining to other items in a given multicomponent measuring instrument. This is a situation that is oftentimes faced by scholars in the behavioral, educational, and social disciplines. Thus, the hybrid IRT models discussed in this chapter present a useful means for dealing with such situations in empirical studies.

12 Introduction to multidimensional item response theory and modeling

In the preceding chapters, we dealt with the unidimensional item response theory (UIRT) modeling framework. While UIRT is extensively used in contemporary behavioral, educational, and social research, in many empirical settings and cases, it can exhibit serious limitations. An extension of UIRT that is needed to manage these limitations is accomplished by the multidimensional item response theory (MIRT) modeling approach. Based on the earlier discussion in the book, the present chapter aims at providing an introduction to this complex topic that has been attracting over the past couple of decades increasing interest by methodologists and substantive scholars.

12.1 Limitations of unidimensional item response theory

A considerable body of research accumulated during the last 30 years has addressed the question when the relatively simple UIRT models may be robust to some violations of their assumptions, in particular that of unidimensionality (compare Reckase [2009]). This research is, however, not conclusive and is relatively limited in terms of settings examined. In addition, it has not sufficiently explicated the conditions and extent of robustness of UIRT to lack of unidimensionality, that is, to multidimensionality of the underlying traits, abilities, or constructs evaluated by a given set of items or measuring instrument. Thus, based on that research, it is not possible to make a more general recommendation for wider use of UIRT models, even under relatively limited violations of homogeneity. In particular, with violations of unidimensionality, results obtained with UIRT could be misleading.

In addition to this caution in general terms, some common-sense reflections also suggest that the actual interactions of examined persons with items in instruments of concern may well be more complex in many realistic empirical situations than implied by UIRT models stipulating single underlying latent variables (compare Reckase [2016]). In particular, deeper knowledge of a substantive area where measurement is conducted as well as details about the study design, organization, and execution will frequently lead to important insights into the evaluation process. They may well imply that examined persons tend to involve more than a single trait or ability when an item set or instrument (even presumably unidimensional) is administered to them. The reason is that to achieve content representation and validity as well as limit the potential for construct underrepresentation (for example, Raykov and Marcoulides [2011]), a researcher may tend to include in a test or scale of interest also items, questions, tasks, or problems that may require more than just the single trait that the instrument is purportedly measuring or claimed to be so. This may often be the case when evaluating achievement or aptitude, for example, in the natural sciences. However, just as well, it can be true i) in clinical settings where one is interested in measuring, say, depression or anxiety; ii) in sociology and political science if one is concerned with assessing certain attitudes; iii) in organizational contexts when interested in evaluating professional mastery or proficiency; or iv) in marketing and business research when consumer satisfaction is, say, of concern (for example, Raykov and Pohl [2013]; Raykov and Calantone [2014]).

To accommodate such likely violations of unidimensionality, a researcher may be inclined to hypothesize that studied subjects vary on more than one trait, that presumably evaluated by the used instrument. That is, a scholar may hypothesize that two or more constructs are responsible for subject performance on a set of administered items. Moreover, some of the items may require multiple skills or abilities to begin with to arrive at their correct or true answers. Perhaps one of the most widely known examples is that of evaluating mathematics ability using in part also word problems. In such tests, in addition to arithmetic, reading ability may well be required, especially in early school grades.

This suggests the need for a generalization of UIRT, which should provide a more adequate representation and modeling of the complexity of the latent space underlying a measuring instrument of concern. (The latent space, as indicated in earlier chapters, can be thought of as being spanned by the set of latent variables that the administered item set taps into and in general need not be unidimensional; compare Raykov and Marcoulides [2008].) This generalization is furnished by MIRT. Its chief goal is to describe the interactions of examined persons with the components of given measuring instruments. Their interactions may well require more than one trait or ability to be adequately explained. The main aim of MIRT is thus to describe better and, by implication, to fit better datasets stemming from studied subjects' performance on complex structure instruments or item sets, relative to the use of UIRT alternatively.

12.2   A main methodological principle underlying multidimensional item response theory

A main principle in MIRT, and in particular its applications in empirical behavioral, educational, and social research, is the following. Even though we allow for a vector (set) of more than one trait to explain the interaction of examined persons with a given measuring instrument, MIRT should not be considered a means of capturing all details of this interaction and especially the relatively "small" and perhaps less relevant aspects of it (compare Reckase [2009]). Therefore, it is important to begin work using a MIRT model with the realization that it should not be expected to explain perfectly, or even close to perfectly, this interaction. That is, just as a model need not be perfect or correct to be useful (as related, for instance, to a now classical "proverb" credited to G. E. P.

Box), a MIRT model of interest that is entertained or used for a given item set need not be close to perfect to be useful.

This point emphasizes that MIRT is to be used to model complex reality rather than to model it perfectly (compare De Boeck and Wilson [2015]). That is, moving from UIRT to MIRT is not done because we want close to perfect fit of models to data, which cannot be achieved otherwise with UIRT. Instead, the use of MIRT is pursued because we want to have access to useful and more adequate models of complex phenomena that cannot be obtained within the framework of UIRT (Reckase [2016]). In other words, MIRT models are meant to be simplifications of reality rather than (close to perfect) reflections of it. These need to be better simplifications and more useful ones than those furnished by UIRT models, which are associated with potential oversimplifications in many empirical settings.

12.3 How can we define multidimensional item response theory?

The preceding discussion in this chapter allows us to define MIRT as an applied statistical framework that is concerned with the probability of a particular response (for example, "correct" answer in the binary or binary scored case) on an item in a given set or instrument under consideration. This probability is hypothesized in MIRT to be represented as follows (compare Reckase [2009]):

$$P(R = r|\boldsymbol{\theta}) = h(r, \boldsymbol{\theta}, \boldsymbol{\gamma}) \tag{12.1}$$

In (12.1), there are a few new symbols relative to the notation used in most of the preceding chapters of the book. Specifically, bolding is used to denote a vector, that is, a set or an entity that contains at least two components, elements, or dimensions. Thereby, $\boldsymbol{\theta}$ is the vector or set of underlying latent variables being evaluated (traits, abilities, constructs, dimensions, continua), which are in general assumed to be correlated. Further, R is the random variable representing the response to an item of concern, with a value r of interest; $h(., ., .)$ is an unknown function; and $\boldsymbol{\gamma}$ is a vector of unknown parameters that describe the characteristics of the items. With binary or binary scored items, as assumed in the rest of this chapter, the response coded $r = 1$ ("correct" response) is typically of special interest. However, this need not be in general the case in theoretical discussions or empirical applications of MIRT.

We stress that the MIRT model in (12.1) depends on multiple traits that are captured in $\boldsymbol{\theta}$, the vector of traits, constructs, abilities, or latent dimensions in general that are tapped into by a measuring instrument or item set of concern. That is, the MIRT model is a latent variable model with two or more latent variables. More concretely, this model relates individual differences on them to such differences in the response patterns on a given set of items. We also emphasize that as in UIRT, multiple parameters are involved in a MIRT model as well. These are typically the elements of the vectors $\boldsymbol{\gamma}$ as well as $\boldsymbol{\theta}$, that is, they include the individual trait or ability levels on all latent dimensions. We also mention that the definition of MIRT includes UIRT as a special case. This special case is obtained when the vector of traits evaluated by the instrument consists of a single element, which is the ability, trait, or latent dimension being tapped into by the instrument.

We can thus describe MIRT as an applied statistical discipline that based on (12.1) aims at i) evaluating the number of latent dimensions underlying a set of considered items, that is, the number of components in $\boldsymbol{\theta}$; ii) ascertaining the particular form of the function h; and iii) estimating the parameters in the vectors $\boldsymbol{\gamma}$ and $\boldsymbol{\theta}$. In other words, MIRT is concerned with the proper selection and estimation (evaluation) of

a. the function h;

b. the dimensionality of $\boldsymbol{\theta}$, that is, the number of its components, in addition to their values for the individual studied subjects; and

c. the elements of the vector $\boldsymbol{\gamma}$.

To achieve the aims of MIRT, researchers have used an increasingly popular class of models in recent decades. These models are closely related to the logistic models used earlier in the present book. We turn next to this model class.

12.4 A main class of multidimensional item response theory models

To describe these models, we will briefly revisit the discussion in chapter 3 concerning the classical (linear) factor analysis (CLFA) model. As will be recalled, the CLFA model was defined there as follows (see, for example, Raykov and Marcoulides [2008, 2011]; $k > 1, m \geq 1$):

$$Y_1 = \mu_1 + a_{11}f_1 + a_{12}f_2 + \cdots + a_{1m}f_m + u_1$$
$$Y_2 = \mu_2 + a_{21}f_1 + a_{22}f_2 + \cdots + a_{2m}f_m + u_2$$
$$\vdots$$
$$Y_k = \mu_k + a_{k1}f_1 + a_{k2}f_2 + \cdots + a_{km}f_m + u_k \qquad (12.2)$$

In (12.2), $f_1, f_2, \ldots, f_m$ are the common factors representing shared sources of latent variability in the manifest continuous measures Y_1 through Y_k and $u_1, u_1, \ldots, u_k$ denote the unique factors. (The reason for the use of the notation f for these factors will become clear shortly.) As we indicated in chapter 3, all factors are assumed to have a mean of 0 and variance 1 (unless otherwise indicated). In addition, the unique factors are assumed uncorrelated with the common factors as well as among themselves.

With the CLFA in mind, we arrived in chapter 4 at a nonlinear factor analysis model by using the generalized linear modeling (GLIM) framework for dichotomous items and the logit link function. That model can be represented here as follows (see also Raykov and Marcoulides [2011]):

$$
\mathrm{logit}(\pi_1) = c_1 + a_{11}f_1 + a_{12}f_2 + \cdots + a_{1m}f_m
$$
$$
\mathrm{logit}(\pi_2) = c_2 + a_{21}f_1 + a_{22}f_2 + \cdots + a_{2m}f_m
$$
$$
\vdots
$$
$$
\mathrm{logit}(\pi_k) = c_k + a_{k1}f_1 + a_{k2}f_2 + \cdots + a_{km}f_m \qquad (12.3)
$$

In (12.3), π_j denote for simplicity the probability of "correct" response on the jth item; for convenience, the same notation is used for the loadings and factors as in (12.2) ($j = 1, \ldots, k$; see also chapter 5).

As indicated in earlier chapters, (12.3) are postulated simultaneously for all items in a considered measuring instrument or item set. These model definition equations stipulate in their right-hand sides the common factors as the sources of shared variability and covariability among the items. Equations (12.3) reflect a multivariate setting where the assumed dichotomous responses Y_1 through Y_k are the dependent variables that in general are interrelated (compare Raykov and Marcoulides [2008]). Because (12.3) also defines a nonlinear factor analysis model, as discussed in chapter 4, we are actually dealing here with a multidimensional (multifactor) model whenever the number of factors is at least 2 (that is, if $m \geq 2$). This model is based on the logit link and is of special relevance in the rest of the present chapter.

The reason is that in fact (12.3) outlines a major MIRT model class. This is because the system of equations (12.3) represent the logit of the probability of correct response on the jth item in terms of $m(m > 1)$ underlying traits, abilities, or latent variables (continua). These are the entities denoted by the f's in the right-hand sides of (12.3) to emphasize the factor analysis item response theory link, and we easily realize that we could have denoted them just as well as θ's (compare Takane and de Leeuw [1987] and Raykov and Marcoulides [2016b]). Equations (12.3) give actually an example of a popular MIRT model class, the so-called compensatory MIRT models ($m \geq 2$; see, for example, Reckase [2009]). In these models (with positive a coefficients), a small value of one of the traits measured can be compensated by a large value of another trait for a given item, thus rendering a prespecified logit and probability of a particular (for example, "correct") response on that item.

From (12.3), the probability of "correct" response on an item is readily obtained with straightforward algebra and a minor change in notation for the present purposes, as indicated in the preceding paragraph (see, for example, chapters 4 and 5 for the unidimensional case):

$$
\pi_j = P_j(\boldsymbol{\theta}) = \exp(c_j + \mathbf{a}_j'\boldsymbol{\theta})/\{1 + \exp(c_j + \mathbf{a}_j'\boldsymbol{\theta})\} = 1/[1 + \exp\{-(c_j + \mathbf{a}_j'\boldsymbol{\theta})\}] \quad (12.4)
$$

($j = 1, \ldots, k$). In (12.4), $P_j(\boldsymbol{\theta})$ symbolizes as previously the probability of "correct" response on the jth item as a function now of the ability vector $\boldsymbol{\theta}$ of m traits, constructs, or abilities being evaluated by the set of considered items. [The former unobserved entities can be treated here as the same as the vector of f's in (12.3), as mentioned above.] Further, $\mathbf{a}_j$ is the vector of loadings for the jth item, that is, its weights on the underlying m factors ($m \geq 2$; priming is used to denote transposition in this chapter). The right-hand side of (12.4) may be seen now as "the same" as that of (5.14) in chapter 5 for the multivariate logistic regression model having a single unobserved predictor, with one exception. Specifically, the previous slope parameter and latent variable in (5.14) are now vectors with m components each in (12.4) (compare Cai and Thissen [2015]).

A special case of the MIRT model defined in (12.4) with $m = 2$ is of relevance in the earlier considered example setting of evaluating mathematics ability using word problems [see also (12.3)]. In that setting, the two abilities assessed are i) θ_1 as arithmetic ability [playing the role then of f_1 in (12.3)] and ii) θ_2 as reading ability [playing the role of f_2 in (12.3)]. This special case is obtained directly from (12.4) by taking the logits of the involved probabilities on their left-hand sides:

$$\text{logit}\{P_1(\boldsymbol{\theta})\} = c_1 + a_{11}\theta_1 + a_{12}\theta_2$$
$$\text{logit}\{P_2(\boldsymbol{\theta})\} = c_2 + a_{21}\theta_1 + a_{22}\theta_2$$
$$\vdots$$
$$\text{logit}\{P_k(\boldsymbol{\theta})\} = c_k + a_{k1}\theta_1 + a_{k2}\theta_2 \tag{12.5}$$

In (12.5), $P_j(\boldsymbol{\theta})$ denotes the probability of correct response on the jth item as a function of the two-dimensional ability vector $\boldsymbol{\theta}$ consisting of the abilities θ_1 and θ_2 mentioned in i) and ii) ($j = 1, \ldots, k$).

The compensatory MIRT models are fit to data and estimated using the same general estimation methods as in UIRT. Thereby, the variances and means of each trait are usually set, respectively, equal to 1 and 0 for the single population and single assessment occasion case, which as mentioned is of relevance throughout most of this book (compare chapter 10 for a multipopulation setting). For further details, we refer to Reckase (2009, 2016). We illustrate next the compensatory MIRT model class on numerical data using Stata.

12.5 Fitting multidimensional item response theory models and comparison with unidimensional item response theory models

In this section, we consider a widely applicable MIRT setting in empirical studies and demonstrate how one can use Stata to fit corresponding MIRT models. In a somewhat indirect relation to the earlier indicated research on robustness of UIRT, we will be specifically interested in testing and comparing a UIRT model with a MIRT model. We will be considering both models as rival means for description and explanation of a given dataset that results from a multi-item measuring instrument (test, scale) with binary or binary scored measures (see, for example, Reckase [2009] for multidimensional

polytomous item response theory models). The MIRT model can also be viewed then as
a means of describing violations of the item homogeneity (unidimensionality) hypothesis
that is implemented in the UIRT model.

The approach discussed below will be of particular interest to an empirical scientist
who i) suspects that an item set or measuring instrument may not be unidimensional;
ii) develops testable hypotheses for the forms of its possible homogeneity violations
in terms of additional abilities it taps into; and iii) wishes to examine whether the
extent to which the instrument is not homogeneous, as reflected in those hypotheses
and in comparison with that of unidimensionality, is limited and perhaps negligible.
In particular, with the next modeling approach, we address the issue of how to test
and compare for an item set of concern MIRT models with $m = 2$ underlying latent
dimensions (abilities, traits, constructs) and specific homogeneity violations against a
UIRT model.

12.5.1   Fitting a multidimensional item response theory model

In some empirical settings, based on prior research or existing theory, a scholar may
be willing to advance a hypothesis stipulating the particular higher-dimensional item
response theory model that is of interest to fit as a possible alternative to a UIRT
model for a given dataset from a studied measuring instrument. For example, when
administering a mathematics ability test as mentioned earlier, one may suspect that
certain of its items (problems, tasks, questions) require for successful completion not
only arithmetic ability per se (abstract thinking ability) but also some nontrivial levels
of reading ability. In such cases, the researcher may be able to develop a sufficiently
precise hypothesis about the structure of the overall instrument, which can then be
examined using uni- and multidimensional item response modeling. This activity can
be readily accomplished with the command `gsem` in Stata, which we already used in
chapter 10. To this end, one can fit the models of interest to the available dataset
and, using the likelihood-ratio test (LRT), statistically examine those that are nested as
well as conduct model choice based on widely used information criteria (for example,
chapter 6).

To illustrate, we use an adapted dataset stemming from a mathematics ability study
with $n = 3000$ students (compare De Boeck and Wilson [2004]). (This dataset is found
in the provided file `mathabilitystudy.dta`.) A main model of interest for its $k = 9$ binary items stipulates that they all evaluate a single ability, which may be referred to as
mathematics ability. (The items are denoted `q1` through `q9` in that data file.) Suppose
that some of these items, say, the first five, are suspected of additionally evaluating reading ability because of the inclusion of word problems. (We appreciate Dr. Raciborski's
insight into and instruction on MIRT modeling with Stata for this empirical setting.)
With the preceding discussion in mind, we are interested in evaluating whether i) the
UIRT model denoted M1, and assuming homogeneity of this test of 9 items, has essentially the same data fit as ii) a MIRT model M2 assuming $m = 2$ abilities evaluated with
the test, namely, mathematics ability by all items and reading ability by the first 5 items

as indicated above. If we find this to be the case, and M1 to be preferable to M2 on model comparison grounds, then we could subsequently proceed treating this nine-item measuring instrument as being in effect unidimensional for practical purposes, that is, as being essentially unidimensional (see also Raykov and Marcoulides [Forthcoming]).

Based on the earlier discussion in this chapter, we begin by fitting the two-dimensional model M2. As described above, M2 assumes all items as loading on a "general" factor (presumably mathematics ability, denoted θ_1), with the first five items loading also on a "local" factor (presumably reading ability, denoted θ_2). We accomplish this with the following command:

```
. use http://www.stata-press.com/data/cirtms/mathabilitystudy.dta
(Adapted Data from De Boeck and Wilson (2004))

. gsem(q*<-Theta1) (q1-q5<-Theta2), logit
> var(Theta1@1 Theta2@1) cov(Theta1*Theta2@0)
```

In this Stata modeling command, we note the use of the shorthand q* as a reference to all items. Further, Theta1 designates the underlying math ability (θ_1) as a construct tapped into by all of them. In addition, Theta2 denotes the reading ability (θ_2) that is presumably also evaluated by the first five items. Lastly, the arrows stand for indicating the assumption that the nine items evaluate the former latent continuum and the first five items assess in addition the latter latent dimension. Because the two abilities are not measured directly, it follows that their scales are not determined, as similarly mentioned in earlier discussions in the book. Thus, we fix their variances to 1 and, with (empirical) model identifiability in mind, their covariance to 0.[1] Thereby, to keep the parallel to the preceding modeling context in the book, we request the logistic link to be used in what may be seen as an underlying GLIM framework implementation in the used generalized structural equation modeling (**gsem**) command (for example, *Stata Structural Equation Modeling Reference Manual* [2017b]).

---

1. We stress that this constraint imposes only a lack of linear relationship and does not exclude nonlinear relationships between the mathematics ability and reading ability presumably underlying the set of used items.

The last presented Stata command yields the following output for model M2.

```
Fitting fixed-effects model:
Iteration 0:   log likelihood =    -8035.23
Iteration 1:   log likelihood = -8024.3463
Iteration 2:   log likelihood = -8024.3398
Iteration 3:   log likelihood = -8024.3398
Refining starting values:

Grid node 0:   log likelihood = -7878.7884
Fitting full model:

Iteration 0:   log likelihood = -7878.7884  (not concave)
Iteration 1:   log likelihood = -7789.6251  (not concave)
Iteration 2:   log likelihood = -7780.4849  (not concave)
Iteration 3:   log likelihood = -7769.7642  (not concave)
Iteration 4:   log likelihood = -7767.5104  (not concave)
Iteration 5:   log likelihood = -7767.3641  (not concave)
Iteration 6:   log likelihood = -7767.2965
Iteration 7:   log likelihood = -7767.2229
Iteration 8:   log likelihood = -7767.1652
Iteration 9:   log likelihood =  -7767.149
Iteration 10:  log likelihood = -7767.1458
Iteration 11:  log likelihood = -7767.1458

Generalized structural equation model           Number of obs    =       1,500
Response       : q1
Family         : Bernoulli
Link           : logit

Response       : q2
Family         : Bernoulli
Link           : logit

Response       : q3
Family         : Bernoulli
Link           : logit

Response       : q4
Family         : Bernoulli
Link           : logit

Response       : q5
Family         : Bernoulli
Link           : logit

Response       : q6
Family         : Bernoulli
Link           : logit

Response       : q7
Family         : Bernoulli
Link           : logit

Response       : q8
Family         : Bernoulli
Link           : logit

Response       : q9
Family         : Bernoulli
Link           : logit

Log likelihood = -7767.1458
 ( 1)  [/]var(Theta1) = 1
 ( 2)  [/]var(Theta2) = 1
```

		Coef.	Std. Err.	z	P>|z|	[95% Conf. Interval]	
q1							
	Theta1	1.295202	.1515172	8.55	0.000	.9982343	1.592171
	Theta2	-.0630324	.1735434	-0.36	0.716	-.4031711	.2771063
	_cons	.6162989	.0758003	8.13	0.000	.467733	.7648647
q2							
	Theta1	.7059701	.0914027	7.72	0.000	.5268241	.8851161
	Theta2	.0681647	.1542835	0.44	0.659	-.2342255	.3705549
	_cons	-.0200923	.0575036	-0.35	0.727	-.1327972	.0926126
q3							
	Theta1	.9403252	.1301827	7.22	0.000	.6851718	1.195479
	Theta2	-.2637436	.2222493	-1.19	0.235	-.6993442	.1718569
	_cons	1.517789	.0982095	15.45	0.000	1.325302	1.710276
q4							
	Theta1	.9816668	.5529784	1.78	0.076	-.102151	2.065485
	Theta2	1.405299	1.659969	0.85	0.397	-1.848181	4.658778
	_cons	-.4908052	.2668261	-1.84	0.066	-1.013775	.0321644
q5							
	Theta1	1.039022	.1326843	7.83	0.000	.778966	1.299079
	Theta2	.2602682	.2573171	1.01	0.312	-.244064	.7646005
	_cons	-1.703043	.1037776	-16.41	0.000	-1.906443	-1.499643
q6							
	Theta1	.9552983	.110694	8.63	0.000	.738342	1.172255
	_cons	-.7909093	.0689374	-11.47	0.000	-.9260241	-.6557945
q7							
	Theta1	.5468859	.0894537	6.11	0.000	.3715598	.722212
	_cons	-1.010067	.063833	-15.82	0.000	-1.135178	-.8849568
q8							
	Theta1	1.122667	.1422268	7.89	0.000	.8439077	1.401427
	_cons	1.978222	.1138226	17.38	0.000	1.755133	2.20131
q9							
	Theta1	.6155707	.091506	6.73	0.000	.4362223	.7949191
	_cons	.9421635	.0640346	14.71	0.000	.8166579	1.067669
var(Theta1)		1	(constrained)				
var(Theta2)		1	(constrained)				

We observe from this output that after an initial difficulty, which is not unexpected in a MIRT setting in empirical studies, the numerical optimization procedure stabilizes and converges to the solution presented. [We treat it as trustworthy here because the final 6 iterations out of a total of 12 full model iterations are uneventful in the sense of not being associated with such difficulties. The reference _cons represents what can be viewed as the observed variable intercepts within that underlying GLIM framework; see the *Stata Structural Equation Modeling Reference Manual* [2017b] and (12.4).]

With this more general model M2 having been fit, we are now ready to compare it with the simpler model. The latter is the UIRT model of concern in this section as well, which we have denoted M1 above. This type of model comparison will be of main interest to empirical scholars who are interested in evaluating the (possible) evidence in a given dataset in favor of i) a MIRT model, like M2 above; or ii) a model of essential unidimensionality and thus in effect a UIRT model, like M1, for an item set or instrument under consideration (compare Reckase [2009]; see Raykov and Marcoulides [Forthcoming]; see also Raykov and Pohl [2013]). We stress that this model comparison is feasible with the present approach when the specific form of M2 is known, expected, suspected, or hypothesized and is implemented in that MIRT model fit as discussed above in this subsection.

12.5.2 Comparing a multidimensional model with an unidimensional model

The UIRT model similarly of concern to us here, M1, stipulates homogeneity of all nine items as indicated above. That is, M1 posits that these items are all assessing a single ability. We fit this model in the following way (to ensure comparability of its fit with that of the MIRT model fit above, as well as applicability of the LRT, we request again the logit link and use otherwise the software default for model identification; see, for example, Raykov and Marcoulides [2011]):

```
. gsem(q*<-Theta1), logit
Fitting fixed-effects model:

Iteration 0:    log likelihood =    -8035.23
Iteration 1:    log likelihood = -8024.3463
Iteration 2:    log likelihood = -8024.3398
Iteration 3:    log likelihood = -8024.3398

Refining starting values:

Grid node 0:    log likelihood = -7813.6654

Fitting full model:

Iteration 0:    log likelihood = -7813.6654  (not concave)
Iteration 1:    log likelihood = -7774.6244
Iteration 2:    log likelihood = -7771.0602
Iteration 3:    log likelihood = -7770.5374
Iteration 4:    log likelihood = -7770.4677
Iteration 5:    log likelihood = -7770.4669
Iteration 6:    log likelihood = -7770.4669
```

```
Generalized structural equation model        Number of obs     =       1,500

Response      : q1
Family        : Bernoulli
Link          : logit

Response      : q2
Family        : Bernoulli
Link          : logit

Response      : q3
Family        : Bernoulli
Link          : logit

Response      : q4
Family        : Bernoulli
Link          : logit

Response      : q5
Family        : Bernoulli
Link          : logit

Response      : q6
Family        : Bernoulli
Link          : logit

Response      : q7
Family        : Bernoulli
Link          : logit

Response      : q8
Family        : Bernoulli
Link          : logit

Response      : q9
Family        : Bernoulli
Link          : logit

Log likelihood = -7770.4669

 ( 1)   [q1]Theta1 = 1
```

		Coef.	Std. Err.	z	P>|z|	[95% Conf. Interval]	
q1							
	Theta1	1	(constrained)				
	_cons	.611308	.0741303	8.25	0.000	.4660152	.7566007
q2							
	Theta1	.5663851	.0955886	5.93	0.000	.3790349	.7537353
	_cons	-.0201098	.0576641	-0.35	0.727	-.1331294	.0929098
q3							
	Theta1	.6878272	.1186269	5.80	0.000	.4553227	.9203318
	_cons	1.474908	.0834104	17.68	0.000	1.311427	1.63839
q4							
	Theta1	.5756396	.0986878	5.83	0.000	.382215	.7690643
	_cons	-.3635994	.059102	-6.15	0.000	-.4794371	-.2477617
q5							
	Theta1	.8529887	.1465932	5.82	0.000	.5656713	1.140306
	_cons	-1.709424	.0997622	-17.13	0.000	-1.904954	-1.513893

q6							
Theta1	.7519696	.1188188	6.33	0.000	.5190889	.9848502
_cons	-.7909196	.0688203	-11.49	0.000	-.9258049	-.6560343
q7							
Theta1	.4309864	.0857326	5.03	0.000	.2629536	.5990191
_cons	-1.010241	.0638023	-15.83	0.000	-1.135291	-.8851908
q8							
Theta1	.8847074	.154192	5.74	0.000	.5824967	1.186918
_cons	1.979141	.1141142	17.34	0.000	1.755481	2.202801
q9							
Theta1	.4830067	.0895953	5.39	0.000	.3074031	.6586104
_cons	.9417878	.0639821	14.72	0.000	.8163853	1.06719
var(Theta1)	1.615295	.3549872			1.049993	2.484947

We now wish to test against each other the nested models fit in this section, M1 and M2, as mentioned earlier. To this end, we use the LRT and model-selection approach used on many occasions in the book (see details in chapters 6 and 7). In the multidimensional setting considered here, and more generally in MIRT contexts where particular homogeneity violations are hypothesized (or are the only ones of concern), the LRT provides a readily applicable means of testing for violations of unidimensionality in an item set or measuring instrument of concern. When no such violations are found, and the simpler model is preferred as well as plausible, this approach offers what may possibly be interpreted as (statistical) "support" for the use of UIRT as an alternative to a rival MIRT model for the instrument in question.

To achieve the above nested model testing, we store next the results obtained for the fit UIRT model M1 in an object called `m1dim` using as before the following Stata command:

```
. estimate store m1dim
```

We will be in a position to conduct then the desired model testing if we `quietly` fit next the more general model M2, store it in a corresponding object, and request in the end the LRT:

```
. quietly gsem(q*<-Theta1) (q1-q5<-Theta2), logit
> var(Theta1@1 Theta2@1) cov(Theta1*Theta2@0)
. estimate store m2dim
. lrtest m2dim m1dim
```

```
Likelihood-ratio test                          LR chi2(5)   =      6.64
(Assumption: m1dim nested in m2dim)            Prob > chi2  =    0.2486
```

Note: The reported degrees of freedom assumes the null hypothesis is not on the boundary of the parameter space. If this is not true, then the reported test is conservative.

The LRT is thus associated with a nonsignificant result. Hence, it suggests that the fit to the analyzed data of the UIRT model, M1, is essentially the same as that of the

MIRT model M2. (The note following the last presented Stata output is not applicable for the pair of nested models fit here, because the nesting restrictions rendering M1 from M2 are not associated with or yield boundary parameter values.)

Using the command `estat ic` after each model is fit, as earlier in the book, we obtain also their Akaike information criterion (AIC) and Bayesian information criterion (BIC) indices, which similarly permit model selection, because M1 and M2 are fit to the same dataset from the same observed variables (see table 12.1).

Table 12.1. Akaike information criterion and Bayesian information criterion indices for two models fit to the used nine-item dataset

Model	AIC	BIC
M1	15576.93	15672.57
M2	15580.29	15702.50

Table 12.1 shows that model M1 is associated with notably lower AIC and in particular BIC indices than model M2 (for example, Raftery [1995]). Hence, the unidimensional model M1 is preferred to M2 also on grounds of this information criteria comparison.

We therefore conclude that there is no evidence in the analyzed data that warrants preferring the MIRT model M2 to the UIRT model M1. In other words, the above findings in fact lend overall support for selecting the unidimensional model over the MIRT model M2. Based on the LRT and information criteria results discussed, the following interpretation is suggested: the extent to which the studied nine-item instrument is not homogeneous, as reflected in model M2, is limited and possibly negligible (in particular as far as empirical or pragmatic purposes are concerned).

We conclude this section by pointing out that the testing and model-selection approach used in it can in fact be more generally used with any prespecified MIRT model when it is of interest to test and compare it against a UIRT model for a considered multicomponent measuring instrument with binary or ordinal items. (See also Raykov and Marcoulides [Forthcoming] for point and interval estimation of a proposed index of essential unidimensionality that describes the extent of violation of unidimensionality in a given item set.)

12.6   Chapter conclusion

In this chapter, we provided an introduction to MIRT. MIRT is a field that is substantially more complex than that of UIRT and has been attracting increased interest among methodologists and substantive researchers over the past couple of decades or so. We considered in particular compensatory MIRT models within the conceptually facilitating context of multivariate multiple logistic regression with unobserved predictors. In addition, we addressed the theoretically and empirically important question as to when UIRT models may be preferable to certain MIRT models. To this end, we have discussed and illustrated an application of the widely used LRT approach for testing a UIRT model against particular MIRT models, aided by information criteria comparisons. This procedure can be readily used, for instance, in empirical behavioral, educational, and social research for exploring whether additional subabilities may be tapped into by some items in an instrument aimed at evaluating an overall construct common to all items. With this feature, the discussed method represents a clearly useful approach that can be recommended for testing specific violations of unidimensionality, or essential unidimensionality, as represented by corresponding MIRT models for studied sets of binary or binary scored items. In addition, the general model-comparison approach underlying the last section is just as applicable in empirical settings involving measuring instruments or sets consisting of or including ordinal items.

13   Epilogue

Item response theory (IRT) and item response modeling (IRM) have enjoyed impressive popularity over the past several decades in the educational, behavioral, and social sciences and well beyond them. With this book, we aimed to provide an introductory to intermediate level treatment of this dynamic and rich measurement field, including relatively brief discussions of some more advanced issues. Another goal of the book was also to illustrate the use of the widely circulated, general purpose statistical software Stata for IRT modeling.

We started with a discussion of the important and consequential links of IRT to other applied statistics areas. These were classical test theory (CTT), factor analysis, and generalized linear modeling—including in particular logistic regression. We see in the emphasis on these highly beneficial connections a characteristic feature of the book. Further, we conceptualized and developed this book with the aim to make it distinct from a number of other treatments of IRT that have appeared both in the past as well as more recently by its lack of juxtaposition of CTT to IRT (see chapter 3 for pertinent details, in particular on CTT). In our opinion, any discussion of IRT that is based on devaluing CTT is in the end a misguided effort that leads to compartmentalization of IRT and IRM with adverse effects (see also footnote 3 to chapter 3). Rather, the book underscored the conceptual and practical benefits resulting from capitalizing on the interrelations that exist between IRT and CTT (when properly used), also as part of the more comprehensive statistical modeling framework of latent variable modeling (see also Raykov and Marcoulides [2016b]). This was exemplified, for instance, with our discussion of the relation between the key concept of the item characteristic curve in IRT in the widely used setting of homogeneous binary or binary scored items on the one hand and the basic concept of the true score from CTT on the other.

We also aimed to offer with the present book such a discussion of main concepts of IRT and their relationships, which provides a coherent and comprehensive basis for the reader to move on subsequently to more specialized treatments and topics within IRT. These include mixture IRT, multilevel IRT, Bayesian IRT, IRT models with covariates, and advanced multidimensional IRT (including multidimensional IRT modeling with polytomous items). Space limitations did not allow us to discuss these and other related parts of the rich IRT and IRM field. However, in our opinion, the foundation needed for approaching these more advanced topics is well explained in the book.

Apart from the last section of chapter 6, throughout the book, we assumed that we were not dealing with missing data or with data that are hierarchical in nature. Missing data and hierarchical settings complicate substantially the use of measurement concepts and relationships, as relevant for IRT and its empirical applications. We point out, how-ever, that under the popular assumption of data missing at random in contemporary behavioral and social research, the modeling approach and procedures discussed in the book can be used also with incomplete datasets. In particular, the `irt` command of Stata that has been used throughout the book offers a straightforward process of fitting popular IRT models via maximum likelihood in this setting as well. In addition, includ-ing "predictors" (explanatory variables) for the latent trait or ability being evaluated by a given set of items or measuring instrument, which are related to the items with missing data, may be beneficial in enhancing the plausibility of the missing at random assump-tion in an empirical study when using IRT model extensions with covariates (compare Little and Rubin [2002]; see also Bock and Moustaki [2007] and Enders [2010]).

In the end, we hope that we have succeeded in showing with this book how use-ful it is to adopt a broader and inclusive, rather than exclusive, view of the relations between IRT and other applied statistics frameworks, particularly those subsumed un-der latent variable modeling, including in particular factor analysis and CTT. We are similarly hopeful that this way of thinking about IRT and IRM will prove to be helpful to many readers open to such an inclusive view. We also hope to have demonstrated how user friendly the application of the popular statistical package Stata is for IRT and IRM purposes. We conclude with the anticipation that the IRT and IRM treatment of-fered in this book will also be helpful to the readers in their subsequent journey into more advanced topics of the impressively wide field of measurement and testing in the behavioral, educational, and social sciences.

References

Agresti, A. 2013. *Categorical Data Analysis*. 3rd ed. Hoboken, NJ: Wiley.

Agresti, A., and B. Finlay. 2009. *Statistical Methods for the Social Sciences*. 4th ed. Upper Saddle River, NJ: Prentice Hall.

Albert, J. H. 2016. Logit, probit, and other response functions. In *Handbook of Item Response Theory, Volume Two: Statistical Tools*, ed. W. J. van der Linden, 3–22. Boca Raton, FL: CRC Press.

Allen, M. J., and W. M. Yen. 1979. *Introduction to Measurement Theory*. Long Grove, IL: Waveland Press.

Andrich, D. 1978. A rating formulation for ordered response categories. *Psychometrika* 43: 561–573.

Baker, F. B., and S.-H. Kim, ed. 2004. *Item Response Theory: Parameter Estimation Techniques*. 2nd ed. Boca Raton, FL: CRC Press.

Bartholomew, D. J. 1996. *The Statistical Approach to Social Measurement*. New York: Academic Press.

Bartholomew, D. J., M. Knott, and I. Moustaki. 2011. *Latent Variable Models and Factor Analysis: A Unified Approach*. 3rd ed. New York: Wiley.

Benjamini, Y., and Y. Hochberg. 1995. Controlling the false discovery rate: A practical and powerful approach to multiple testing. *Journal of the Royal Statistical Society, Series B* 57: 289–300.

Benjamini, Y., and D. Yekutieli. 2001. The control of the false discovery rate in multiple testing under dependency. *Annals of Statistics* 29: 1165–1188.

Birnbaum, A. 1968. Some latent trait models. In *Statistical Theories of Mental Test Scores*, ed. F. M. Lord and M. R. Novick, 397–424. Reading, MA: Addison–Wesley.

Bock, R. D. 1972. Estimating item parameters and latent ability when responses are scored in two or more nominal categories. *Psychometrika* 37: 29–51.

———. 1997. A brief history of item theory response. *Educational Measurement: Issues and Practice* 16: 21–33.

Bock, R. D., and M. Aitkin. 1981. Marginal maximum likelihood estimation of item parameters: Application of an EM algorithm. *Psychometrika* 46: 443–459.

Bock, R. D., and I. Moustaki. 2007. Item response theory in a general framework. In *Handbook of Statistics 26: Psychometrics*, ed. C. R. Rao and S. Sinharay, 469–513. Oxford: Elsevier.

Bollen, K. A. 1989. *Structural Equations with Latent Variables*. New York: Wiley.

Cai, L., and D. Thissen. 2015. Modern approaches to parameter estimation in item response theory. In *Handbook of Item Response Theory Modeling: Applications to Typical Performance Assessment*, ed. S. P. Reise and D. A. Revicki, 41–59. New York: Taylor & Francis.

Camilli, G. 1994. Origin of the scaling constant d = 1.7 in item response theory. *Journal of Educational and Behavioral Statistics* 19: 293–295.

Casella, G., and R. L. Berger. 2002. *Statistical Inference*. 2nd ed. Pacific Grove, CA: Duxbury.

de Ayala, R. J. 2009. *The Theory and Practice of Item Response Theory*. New York: Guilford Press.

De Boeck, P., and M. Wilson. 2004. A framework for item response models. In *Explanatory Item Response Models: A Generalized Linear and Nonlinear Approach*, ed. P. De Boeck and M. Wilson, 3–42. New York: Springer.

———. 2015. Multidimensional explanatory item response modeling. In *Handbook of Item Response Theory Modeling: Applications to Typical Performance Assessment*, ed. S. P. Reise and D. A. Revicki, 252–271. New York: Taylor & Francis.

Dobson, A. J., and A. G. Barnett. 2008. *An Introduction to Generalized Linear Models*. 3rd ed. Boca Raton, FL: CRC Press.

Enders, C. K. 2010. *Applied Missing Data Analysis*. New York: Guilford Press.

Finney, D. J. 1947. *Probit Analysis: A Statistical Treatment of the Sigmoid Response Curve*. Cambridge: Cambridge University Press.

Goldstein, H. 2011. *Multilevel Statistical Models*. 4th ed. London: Arnold.

Goldstein, H., and M. J. R. Healy. 1995. The graphical presentation of a collection of means. *Journal of the Royal Statistical Society, Series A* 158: 175–177.

Graham, J. W. 2009. Missing data analysis: Making it work in the real world. *Annual Review of Psychology* 60: 549–576.

Haley, D. C. 1952. Estimation of the dosage mortality relationship when the dose is subject to error. Technical Report 15, Applied Mathematics and Statistics Laboratory, Stanford University. https://statistics.stanford.edu/sites/default/files/SOL%20ONR%2015.pdf.

Hambleton, R. K., and H. Swaminathan. 1985. *Item Response Theory: Principles and Applications*. Boston, MA: Kluwer.

Hambleton, R. K., H. Swaminathan, and H. J. Rogers. 1991. *Fundamentals of Item Response Theory*. Thousand Oaks, CA: Sage.

Harman, H. H. 1976. *Modern Factor Analysis*. 3rd ed. Chicago: University of Chicago Press.

Holland, P. W., and H. Wainer, ed. 1993. *Differential Item Functioning*. Hillsdale, NJ: Lawrence Erlbaum.

Hosmer, D. W., Jr., S. Lemeshow, and R. X. Sturdivant. 2013. *Applied Logistic Regression*. 3rd ed. Hoboken, NJ: Wiley.

Hox, J. J. 2010. *Multilevel Analysis: Techniques and Applications*. 2nd ed. New York: Taylor & Francis.

Johnson, M. S., and S. Sinharay. 2016. Bayesian estimation. In *Handbook of Item Response Theory, Volume Two: Statistical Tools*, ed. W. J. van der Linden, 237–258. Boca Raton, FL: CRC Press.

Johnson, N. L., S. Kotz, and N. Balakrishnan. 1995. *Continuous Univariate Distributions, Volume 2*. 2nd ed. New York: Wiley.

Johnson, R. A., and D. W. Wichern. 2007. *Applied Multivariate Statistical Analysis*. 6th ed. Upper Saddle River, NJ: Prentice Hall.

Jöreskog, K. G. 1971. Statistical analysis of sets of congeneric tests. *Psychometrika* 36: 109–133.

Kamata, A., and D. J. Bauer. 2008. A note on the relation between factor analytic and item response theory models. *Structural Equation Modeling: A Multidisciplinary Journal* 15: 136–153.

Kohli, N., J. Koran, and L. Henn. 2015. Relationships among classical test theory and item response theory frameworks via factor analytic models. *Educational and Psychological Measurement* 75: 389–405.

Lawley, D. N. 1943. On problems connected with item selection and test construction. *Proceedings of the Royal Society of Edinburgh, Section A* 61: 273–287.

———. 1944. The factorial analysis of multiple item tests. *Proceedings of the Royal Society of Edinburgh, Section A* 62: 74–82.

Little, R. J. A., and D. B. Rubin. 2002. *Statistical Analysis with Missing Data*. 2nd ed. Hoboken, NJ: Wiley.

Lord, F. M. 1952. A theory of test scores. Psychometric Monograph, No. 7.

———. 1953. An application of confidence intervals and of maximum likelihood to the estimation of an examinee's ability. *Psychometrika* 18: 57–76.

———. 1977. Practical applications of item characteristic curve theory. *Journal of Educational Measurement* 14: 117–138.

———. 1980. *Applications of Item Response Theory to Practical Testing Problems.* Mahwah, NJ: Lawrence Erlbaum.

Lord, F. M., and M. R. Novick, ed. 1968. *Statistical Theories of Mental Test Scores.* Reading, MA: Addison–Wesley.

Lubke, G., and B. O. Muthén. 2005. Investigating population heterogeneity with factor mixture models. *Psychological Methods* 10: 21–39.

Masters, G. N. 1982. A Rasch model for partial credit scoring. *Psychometrika* 47: 149–174.

———. 2016. Partial credit model. In *Handbook of Item Response Theory, Volume One: Models*, ed. W. J. van der Linden, 109–126. Boca Raton, FL: CRC Press.

McDonald, R. P. 1967. Nonlinear factor analysis. Psychometric Monograph, No. 15.

———. 1999. *Test Theory: A Unified Treatment.* Mahwah, NJ: Lawrence Erlbaum.

Mellenbergh, G. J. 1994. Generalized linear item response theory. *Psychological Bulletin* 115: 300–307.

Meredith, W. 1993. Measurement invariance, factor analysis and factorial invariance. *Psychometrika* 58: 525–543.

Messick, S. 1995. Validity of psychological assessment: Validation of inferences from persons' responses and performances as scientific inquiry into score meaning. *American Psychologist* 50: 741–749.

Michell, J. 2005. *Measurement in Psychology: A Critical History of a Methodological Concept.* Cambridge: Cambridge University Press.

Mislevy, R., and R. D. Bock. 1984. *BILOGI maximum likelihood item analysis and test scoring: Logistic model.* Mooresville, ID: Scientific Software International.

Mulaik, S. A. 2009. *Foundations of Factor Analysis.* 2nd ed. Boca Raton, FL: CRC Press.

Muraki, E. 1992. A generalized partial credit model: Application of an EM algorithm. *Applied Psycholigical Measurement* 16: 159–176.

Muthén, B. O. 1984. A general structural equation model with dichotomous, ordered categorical, and continuous latent variable indicators. *Psychometrika* 49: 115–132.

———. 2002. Beyond SEM: General latent variable modeling. *Behaviormetrika* 29: 81–117.

Muthén, B. O., and T. Asparouhov. 2016. Multidimensional, multilevel, and multi-timepoint item response modeling. In *Handbook of Item Response Theory, Volume One: Models*, ed. W. J. van der Linden, 527–540. Boca Raton, FL: CRC Press.

Nelder, J. A., and R. W. M. Wedderburn. 1972. Generalized linear models. *Journal of the Royal Statistical Society, Series A* 135: 370–384.

Nering, M. L., and R. Ostini, ed. 2010. *Handbook of Polytomous Item Response Theory Models*. New York: Taylor & Francis.

Osterlind, S. J., and H. T. Everson. 2009. *Differential Item Functioning*. 2nd ed. Thousand Oaks, CA: Sage.

Rabe-Hesketh, S., and A. Skrondal. 2016. Generalized linear latent and mixed modeling. In *Handbook of Item Response Theory, Volume One: Models*, ed. W. J. van der Linden, 503–526. Boca Raton, FL: CRC Press.

Raftery, A. E. 1995. Bayesian model selection in social research. *Sociological Methodology* 25: 111–163.

Raykov, T. 2011. On testability of missing data mechanisms in incomplete data sets. *Structural Equation Modeling: A Multidisciplinary Journal* 18: 419–430.

Raykov, T., and R. J. Calantone. 2014. The utility of item response modeling in marketing research. *Journal of the Academy of Marketing Science* 42: 337–360.

Raykov, T., D. M. Dimitrov, G. A. Marcoulides, T. Li, and N. Menold. Forthcoming. Examining measurement invariance and differential item functioning with discrete latent construct indicators: A note on a multiple testing procedure. *Educational and Psychological Measurement*.

Raykov, T., P. A. Lichtenberg, and D. Paulson. 2012. Examining the missing completely at random mechanism in incomplete data sets: A multiple testing approach. *Structural Equation Modeling: A Multidisciplinary Journal* 19: 399–408.

Raykov, T., and G. A. Marcoulides. 2006. *A First Course in Structural Equation Modeling*. 2nd ed. Mahwah, NJ: Lawrence Erlbaum.

———. 2008. *An Introduction to Applied Multivariate Analysis*. New York: Taylor & Francis.

———. 2011. *Introduction to Psychometric Theory*. New York: Taylor & Francis.

———. 2013. *Basic Statistics: An Introduction with R*. Lanham, MD: Rowman & Littlefield.

———. 2016a. On examining specificity in latent construct indicators. *Structural Equation Modeling: A Multidisciplinary Journal* 23: 845–855.

————. 2016b. On the relationship between classical test theory and item response theory: From one to the other and back. *Educational and Psychological Measurement* 76: 325–338.

————. Forthcoming. On studying common factor dominance and approximate unidimensionality in multi-component measuring instruments with discrete items. *Educational and Psychological Measurement*.

Raykov, T., G. A. Marcoulides, and C. Chang. 2016. Examining population heterogeneity in finite mixture settings using latent variable modeling. *Structural Equation Modeling: A Multidisciplinary Journal* 23: 726–730.

Raykov, T., G. A. Marcoulides, S. Gabler, and Y. Lee. 2017. Testing criterion correlations with scale component measurement errors using latent variable modeling. *Structural Equation Modeling: A Multidisciplinary Journal* 24: 468–474.

Raykov, T., G. A. Marcoulides, and T. Patelis. 2015. The importance of the assumption of uncorrelated errors in psychometric theory. *Educational and Psychological Measurement* 75: 634–647.

Raykov, T., and S. Pohl. 2013. On studying common factor variance in multiple-component measuring instruments. *Educational and Psychological Measurement* 73: 191–209.

Reckase, M. D. 2009. *Multidimensional Item Response Theory*. New York: Springer.

————. 2016. Logistic multidimensional models. In *Handbook of Item Response Theory, Volume One: Models*, ed. W. J. van der Linden, 189–210. Boca Raton, FL: CRC Press.

Rizopoulos, D. 2007. ltm: An R package for latent variable modeling and item response theory analyses. *Journal of Statistical Software* 17(5): 1–25.

Roussas, G. G. 1997. *A Course in Mathematical Statistics*. 2nd ed. San Diego: Academic Press.

Samejima, F. 1969. Estimation of latent ability using a response pattern of graded scores. Psychometric Monograph No. 17, Psychometric Society, Richmond, VA.

————. 2016. Graded response models. In *Handbook of Item Response Theory, Volume One: Models*, ed. W. J. van der Linden, 95–108. Boca Raton, FL: CRC Press.

Sijtsma, K., and I. W. Molenaar. 2002. *Introduction to Nonparametric Item Response Theory, Volume 5*. Thousand Oaks, CA: Sage.

Skrondal, A., and S. Rabe-Hesketh. 2004. *Generalized Latent Variable Modeling: Multilevel, Longitudinal, and Structural Equation Models*. Boca Raton, FL: Chapman & Hall/CRC.

Spearman, C. 1904a. "General intelligence" objectively determined and measured. *American Journal of Psychology* 15: 201–293.

————. 1904b. The proof and measurement of association between two things. *American Journal of Psychology* 15: 72–101.

StataCorp. 2017a. *Stata 15 Item Response Theory Reference Manual*. College Station, TX: Stata Press.

————. 2017b. *Stata 15 Structural Equation Modeling Reference Manual*. College Station, TX: Stata Press.

Takane, Y., and J. de Leeuw. 1987. On the relationship between item response theory and factor analysis of discretized variables. *Psychometrika* 52: 393–408.

Thissen, D., and L. Steinberg. 1986. A taxonomy of item response models. *Psychometrika* 51: 567–577.

Timm, N. H. 2002. *Applied Multivariate Analysis*. New York: Springer.

van der Linden, W. J., ed. 2016a. *Handbook of Item Response Theory*. Boca Raton, FL: CRC Press.

van der Linden, W. J. 2016b. Unidimensional logistic response models. In *Handbook of Item Response Theory, Volume One: Models*, ed. W. J. van der Linden, 13–30. Boca Raton, FL: CRC Press.

von Davier, M. 2016. Rasch model. In *Handbook of Item Response Theory, Volume One: Models*, ed. W. J. van der Linden, 31–50. Boca Raton, FL: CRC Press.

Wasserman, L. 2004. *All of Statistics: A Concise Course in Statistical Inference*. New York: Springer.

Zimmerman, D. W. 1975. Probability spaces, hilbert spaces, and the axioms of test theory. *Psychometrika* 40: 395–412.

Zorich, V. A. 2016. *Mathematical Analysis II*. 2nd ed. New York: Springer.

Author index

Subject index